AF594536

The Botanists' Library

334.

Asphodelus ramosus.

The Botanists' Library

The Most Important Botanical Books in History

CAROLYN FRY
AND
EMMA WAYLAND

First published in 2024 by Ivy Press
an imprint of The Quarto Group.
One Triptych Place, London, SE1 9SH
United Kingdom
T (0)20 7700 6700
www.Quarto.com

A catalogue record for this book is available from the British Library.

ISBN 978-0-71129-494-3
Ebook ISBN 978-0-71129-495-0

Design by Kevin Knight
Art Director Paileen Currie
Senior Editor Michael Brunström
Additional editorial work Nayima Ali and Izzy Toner
Production Manager Rohana Yusof
Picture research Steve Behan

Printed in China

10 9 8 7 6 5 4 3 2 1

CONTENTS

INTRODUCTION **6**

1 **EARLY TEXTS SHOW EMERGING KNOWLEDGE OF PLANTS** (ANCIENT TIMES–1450) **12**

2 **THE COMING OF PRINT** (1450–1600) **50**

3 **BOTANISTS STRIVE TO KNOW AND CLASSIFY MORE PLANTS** (1600–1750) **86**

4 **THE GLOBAL AND THE LOCAL** (1750–1830) **138**

5 **BOTANY BECOMES A SCIENCE** (1830–1950) **184**

6 **THE MODERN BOTANISTS' LIBRARY TAKES SHAPE** (1950–THE PRESENT DAY) **228**

INDEX **264**
FURTHER READING **270**
PICTURE CREDITS **271**
ACKNOWLEDGEMENTS **272**

INTRODUCTION

What books should be included in the discerning botanist's library? It's a difficult question to answer given that our relationship with plants extends back thousands of years and botany is still evolving as a field of study today. The subject is diverse in its scope, too, with the expansion of botanical knowledge over time linked to medicine, myth, religion, agriculture, art, exploration, horticulture, politics, war, the environment, technology and more. And it is globally relevant. As has been the case down the centuries, we all depend on plants for food, materials and the 'ecosystem services' (such as clean water and the air we breathe) that biodiversity as a whole provides. In this book, we define botany loosely as the study of plants (allied to that of fungi), and largely focus on how the field emerged and subsequently developed into modern plant science.

Our story begins with the earliest manuscripts featuring plants. Some of these record information dating back even further, either from older manuscripts now no longer in existence or from when knowledge was passed down the generations by word of mouth. Although the oldest accessions in our library date back 3,500 years, these are not the earliest evidence of humans recording their relationship with plants: in modern-day Brazil's Serra da Capivara National Park, depicted in red paint in pictograms, is a 10,000-year-old image of people worshipping a tree. However, this painting's location on the wall of a cave disqualifies it from inclusion on our list. One rule we have applied when deciding which works to include is that they must be easily portable, and therefore capable of being brought to our library for safekeeping. For the most part, our collection contains physical manuscripts and books, but a few digital resources have also crept in, in a nod to the future trajectory of plant science.

Of course, some decisions about what to include have been made for us. Books are easily destroyed by pests, fire, damp and other unfavourable conditions, and plenty have succumbed to such a fate over the millennia. Others have survived more by luck than judgement – as could be said for ancient Greek botanist Theophrastus' important works *Historia Plantarum (Enquiry into Plants)* and *De causis plantarum (Causes of Plants)*. According to the historian and geographer Strabo, writing at around the start of the first millennia CE, Theophrastus bequeathed his library to his disciple Neleus, who took it to Scepsis (in modern-day Turkey). Neleus' heirs hid the collection in a cellar to prevent it from being confiscated by ransacking kings but, instead, it was attacked by moths and moisture. Fortunately – and as is the case for many ancient documents – subsequent duplication and translation kept Theophrastus' work alive down the centuries for inclusion in our collection.

Fire has been a frequent curse in the history of libraries, responsible for the destruction of many texts that, under other circumstances, could have informed the future. One of the worst such events took place in 1193, at the renowned university at Nalanda, in Bihar, India. Almost its entire library of nine million palm-leaf manuscripts went up in flames at the hands of Turko-Afghan invaders. Established during the Gupta Empire and operating from 427 CE, this Buddhist monastic centre of learning had taught, among many other

LEFT

De materia medica, Dioscorides, **(60–70 CE), Arabic (Iraqi School) translation of 1334**

Plants have been used medicinally around the world for millennia. Human-like mandrakes, such as those seen here, were once associated with fertility and used as an aphrodisiac.

RIGHT

Les Vélins du Roi,
Nicolas Robert,
(*c.* 1665–85)

Botanical illustrations in historical texts can help to unravel how plant species have moved around the world. The sunflower seen here was introduced to Europe from North America by the Spanish in the 16th century.

subjects, the nature-based healing methods of Ayurveda. Only a few of the manuscripts survived the fire. We can only guess at the knowledge of plants and their medicinal uses that was lost in the destroyed texts.

Botany has been linked to conflict at many times during its long history. Today, there are strong controls over the movement of plants around the world, and global agreements help ensure that any benefits arising from research and development involving plants feed back into the country from where the material originated. It was not always that way. Starting from the early Middle Ages, European countries established trading posts and colonized lands in Asia, Africa and the Americas, frequently exploiting the local natural resources. Many botanical texts were made, describing – for the benefit of European nations – interesting or potentially useful plant materials seen growing in foreign soils. Often beautifully illustrated, these works are important today for the record they provide of colonial history and its intrinsic links with botany, as well as of past biodiversity. However, while enlightening us as to the colonial mindset, they also serve to silence the indigenous experience.

BELOW

***Florentine Codex*, Bernardino de Sahagún, (16th century)**

This work documents the natural history and culture of the Aztecs, including edible plants encountered by its Spanish author.

Botanical books made by colonizing powers incorporated the knowledge of local experts, often without acknowledging them, some of whom were enslaved people. As well as highlighting these cases, we have tried to show where efforts are being made today to redress such past actions. An example is the ongoing work to identify the individual talented Indian artists who were commissioned to paint plants by the British East India Company in the eighteenth and nineteenth centuries, but whose names and diverse styles were rendered invisible by the use of the blanket term 'Company School' or '*Kampani kalam*' to describe these works.

One of the enduring names connected with botany is Carl Linnaeus, the eighteenth-century Swedish naturalist who devised the first classification system based on plants' sexual reproduction, and introduced the Latin binomial nomenclature still used to name species today. This effort, informed by 3,000 letters Linnaeus received from 60 correspondents in Europe, America, Asia and Africa, had far-reaching consequences for botany – hence the inclusion of his books in our library. We acknowledge that, in also classifying humans into four varieties – European white, American reddish, Asian tawny and African black – Linnaeus also initiated 'scientific racism', which uses pseudoscience to justify racism. As this field grew it was used to argue the superiority of Europeans versus non-Europeans, and to justify actions such as slavery and genocide.

Prejudice against women also features strongly in botany's history. Although we have tried to highlight works by women, the majority of books presented in the six chapters of the *Botanists' Library* are by men. British writer and artist Beatrix Potter's story gives an insight into the struggle faced by women in the past to make their names in natural sciences. Potter is best known for her illustrated children's books, which have sold more than 250 million copies. However, earlier in her life she was a keen natural scientist. In 1897, she put forward a paper to the Linnean Society in London (which houses Linnaeus' collections), but as a woman was not allowed to be a member of the society nor attend the meeting when her paper was read. When the society's members did not pay much attention to her work, and fearing her samples to be contaminated, Potter withdrew her paper, which became lost. Only after Potter left hundreds of mycological artworks to a museum in the Lake District, UK, on her death in 1943, were her scientific talents recognized. In 1997, the Linnean Society offered a posthumous apology to her, acknowledging the sexism displayed towards Potter and other women. Potter's precise and beautiful paintings and drawings of fungi are now helping modern mycologists in their efforts to identify species.

The books and manuscripts included in our *Botanists' Library* reflect our interpretation of how botany has developed over time amid shifting attitudes, approaches, knowledge and resources. Chapter 1 spans ancient times to 1450, covering what the

BELOW

Specimens of the plants and fruits of the island of Cuba, **Anne Wollstencraft, (19th century)**

Female natural scientists are few and far between in the historical botanical record, reflecting past social attitudes.

ancient peoples of Egypt, India and Greece knew about plants; how Greco-Roman knowledge was duplicated for centuries; China's tradition of *Ben Cao* texts; the flourishing of Islamic science between the eighth and thirteenth centuries; and how this revived interest in botany across Europe in the Renaissance. Chapter 2 covers 1450–1600, encompassing the arrival of the printing press in Europe; the emergence of illustrated herbals, botanic gardens and herbaria for classifying plants in Europe; the early classification of plants from the Arabic world; and how Spanish and Portuguese exploration gave rise to the first European herbals from South America. Chapter 3 follows botany's development between 1600 and 1750, covering the emergence of flora and florilegia; botanical works by European colonial powers; early illustrations of plants and animals together; how China and Japan exchanged botanical knowledge with Europe; and Linnaeus' sexual classification and naming system for plants. Chapter 4 takes the story on from 1750 to 1830, covering new approaches to classification; the development of botany in European countries and the expansion of colonialism; and the rise of professional botany in universities. Chapter 5 covers 1830 to 1950, spanning botany's transition to plant science; the development of evolutionary theory; and the emergence of the conservation movement. And finally, Chapter 6 brings our story up to date, with the ground-breaking revelation of the structure of DNA; use of the model plant *Arabidopsis thaliana*; the introduction of genetics-based classification; concerns about planetary health; the ongoing value of botanical illustration; and the books that are educating new scholars and lay people as to the wonder of plants.

Our interpretation of how the study of plants has waxed and waned down the millennia is one of many possible and varied narratives. And the books we have included to tell our story represent but a tiny fraction of the myriad plant-related words written around the world during the timespan of this book. In libraries of the real world, new books come and old books go, so our collection of texts represents a snapshot in time. As to what the library looks like, we leave that entirely up to you. Perhaps you view it as being like the ancient Library of Alexandria in Egypt, with its Greek columns, gardens, *peripatos* (walkways) and inscription reading: 'The place of the cure of the soul'. Alternatively, you may envisage it as a Renaissance-style masterpiece, with elaborate painted ceilings, such as the 1592 library of El Escorial in Madrid, Spain. Or you might find a building like the new Beijing City Library in China more appropriate, for its giant tree-like columns.

Whatever the architecture and ambience of your botanists' library, we welcome you. Come in, make yourself comfortable and lose yourself in the millennia-long journey that has made the study of plants what it is today. It is a tumultuous story of curiosity, power and greed – and of our relentless quest to quantify and classify the natural world that is ongoing today.

Silence please!

EARLY TEXTS SHOW EMERGING KNOWLEDGE OF PLANTS

(ANCIENT TIMES–1450)

OPPOSITE

***Ebers Papyrus*, unknown author(s), (*c.* 1500 BCE)**

This papyrus records ancient Egyptian preparations, spells and incantations for treating ailments.

Plants must have seemed wondrous providers to early humans. Not only did they offer sustenance, healing and shelter, many had beautiful flowers and exquisite perfume. So, it is not surprising that when written texts began to supplement oral traditions of sharing knowledge, from around 3,500 years ago, plants were among the earliest subjects to feature. As the thinkers of their day examined how vegetation around them grew, mused on the variety of plants they saw, sought to understand what beneficial or dangerous qualities plants had, and admired floral colours and scents, they put stylus to palm leaf, reed brush to papyrus and quill to parchment to record their findings. The knowledge that has come down the generations to us from their time tells us that our early ancestors' understanding of plants was closely entwined with food, religion, myth, magic and medicine.

The world explored and recorded in these early texts is one in which their ancestors had only relatively recently switched from hunter-gathering to farming, given up their nomadic lifestyles to settle, and built the first cities. As farming had begun to provide an assured food supply, fertility had risen, driving populations up and setting the scene for the earliest civilizations to emerge: in Mesopotamia (around 3500 BCE); Egypt (3100 BCE); the Indus valley of northwestern South Asia (2500 BCE) and China (1800 BCE). These civilizations each invented their own forms of writing, honing their skills on clay tablets, pottery, bones, ivory or turtle shells. Once they realized that plants and animal skins could provide more portable writing media, the use of the written word widened and the concept of a library came into view.

1 The plants that healed ancient Egyptians

The *Ebers Papyrus*, a scroll more than 18m (60ft) long that today resides in Leipzig University Library, Germany, provides an early insight into botanical knowledge in ancient Egypt. The most extensive and best-preserved record of ancient Egyptian medicine, it was written around 1500 BCE, while Pharaoh Amenhotep I was pushing Egypt's boundaries into Nubia (modern Sudan) and putting his mark on the Temple of Karnak at Thebes (within today's city of Luxor). Written in black and red ink, its 110 pages list 877 prescriptions made up from 328 ingredients mostly derived from plants. These were likely compiled from earlier documents, long since lost, built on the oral recollections of generations of physicians.

The papyrus shows that ancient Egyptian doctors had well-considered views about how the body worked, what caused it to malfunction and which plants to use as medicines. They believed that the body contained '*mtw*', an infinity of intertwined vessels that carried bodily fluids, such as blood and urine. Observing that obstructions to the flow of the River Nile caused floods in some areas and droughts in others, they surmised that blockages within the body's vessels and ducts would result in disease. Seeking inspiration from the natural world, too, to treat these ailments, they applied honey and onions to treat infections; prescribed salicin extracted from the willow tree to reduce pain and inflammation (this chemical was later used to develop aspirin); and used plants including poppies to sedate patients. Opium, which occurs naturally in the latter, might have helped to calm those facing amputation; some mummies from ancient Egyptian tombs have been found with wooden prosthetic toes exhibiting signs of wear and tear.

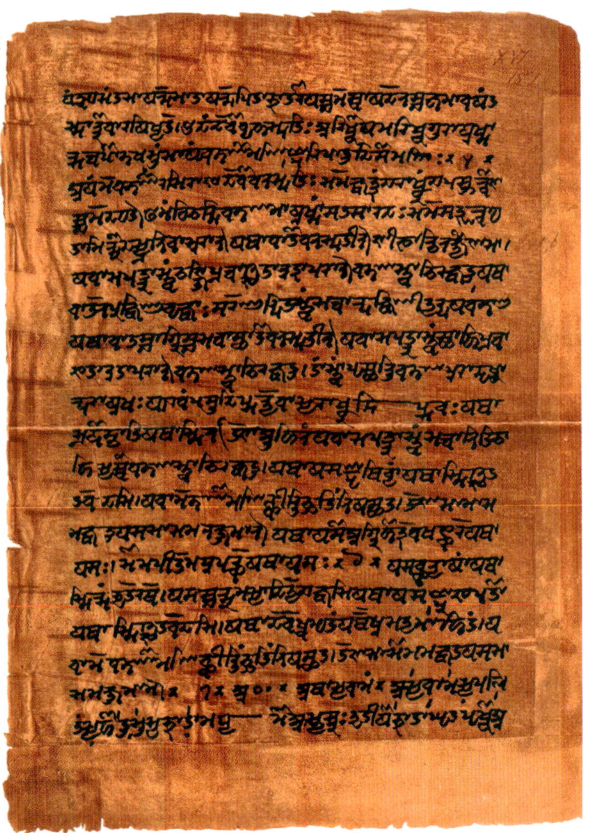

ABOVE

***Atharvaveda*, unknown author(s), (*c.* 1500–1200 BCE)**

This ancient Indian text includes prayers to protect crops from lightning and drought, as well as charms, spells and verses.

The early physicians were aware of cancer, too. The *Ebers Papyrus*' 'Treatise on Tumours' contains the first mention of a soft-tissue tumour, and it possibly also references cancers of the skin, stomach, uterus and rectum. The treatments it prescribes include frankincense, which is the aromatic dried sap of *Boswellia* trees, and barley, to be 'cooked in water, without boiling' and after being taken out of the fire 'mixed with date's cores'. While some treatments may have helped patients, less than a third of 260 medical prescriptions given in the *Hearst Papyrus*, another medical record written at around the same time as the *Ebers Papyrus*, had ingredients likely to be active in treating the condition. Among its prescriptions, it recommends the use of the castor oil plant for expelling fluid, and provides incantations to bring down evil and sickness in the body. Religion and magic were both deeply entwined within ancient Egyptian medicine, with physicians routinely using magical spells and incantations alongside herbal remedies and surgical procedures.

2 The roots of India's Ayurveda

The earliest texts known from India (which frequently reference plants) comprise a vast corpus (*samhitā*) of Sanskrit poems, philosophical dialogue, myth and ritual incantations. Known as the *Vedas*, they were composed between 1500 BCE and 1200 BCE, and comprise: the *Rigveda* (representing the oldest forms of Sanskrit text); *Yajurveda* (ritual-offering mantras and chants); *Samaveda* (melodies and chants) and *Atharvaveda* (magic and charms). These verses represent the oldest scriptures of Hinduism, and are considered to be the work of ancient sages after intense meditation, or of Brahma – one of three major gods of the Hindu faith. The texts show that linking plants to religion was undertaken to protect nature; that the ancient Indians developed the first plant-classification system; and that, like their Egyptian counterparts, the people of Vedic times knew much about the medicinal value of plants.

One verse from the *Yajurveda*, a prayer to a herb being dug up from the soil for medicinal use, evokes the pain felt in the heart on disturbing the status quo of nature. In it, the person uprooting the plant feels guilty and asks the herb to spare him, his wife, sons and cattle from the curse of disease. In a connected verse, the person explains how, in reverence to Mother Earth, he is leaving sufficient plants behind (to allow for regeneration). The *Rigveda* considers plants to be the 'protectors of humanity' and references the *soma* as both a plant (which could be used to make an intoxicating drink) and a god. The identity of this plant has not been proven but the twenty-six candidates proposed thus far by scholars include species from the *Ephedra* and *Cynanchum* genera, and fly agaric (*Amanita muscaria*) mushrooms.

The Vedic people sought to understand and classify the plants they encountered. The *Atharvaveda* notes that the sky is the father and the earth the mother of vegetation, that plants first arose in the sea and that some later adopted a terrestrial way of living. The *Rigveda* recognizes three groups of plants – *vrksha* (trees), *osadhi* (herbs, particularly medicinal ones) and *virudh* (creepers), which it further classifies into flowering, non-flowering, fruit-bearing and non-fruiting plants. The ancient Indians used the word *aushadhi* for plants that were effective in treating diseases. The *Atharvaveda* mentions around fifty diseases and more than 100 plants, as well as minerals and animals. Several hymns are dedicated to praising medicinal herbs.

The text classifies medicinal plants on characteristics such as their colour, how they grow and where they originate. *Ashvattha*, the sacred fig (*Ficus religiosa*), is deemed 'useful for all kinds of diseases'; *darbha,* kusha grass (*Demostachya bipinnata*), is described as an antidote for snake poison; and *rajani*, turmeric (*Curcuma longa*), is considered appropriate for treating both leprosy and baldness. Vedic people took various approaches to healing, including the use of herbs, the chanting of mantras, amulets, touch, cleansing with water and exposure to the sun. As one verse of the *Rigveda* shows, those with the knowledge to treat people were highly valued in society: 'He who hath store of herbs at hand like kings amid a crowd of men – physician is that sage's name, fiend slayer, chaser of disease.' The *Vedas* were written to be transmitted orally, which continues even today. But hard-copy texts of considerable age exist, including *Rigveda* manuscripts written on birch bark and paper dating from 1464, now housed at the Bhandarkar Oriental Research Institute in Pune, India.

BELOW

***Rigveda*,**
unknown author(s),
(*c.* 1500–1200 BCE)

The *Rigveda* is the oldest of four ancient texts from India, which contain frequent references to plants and their uses.

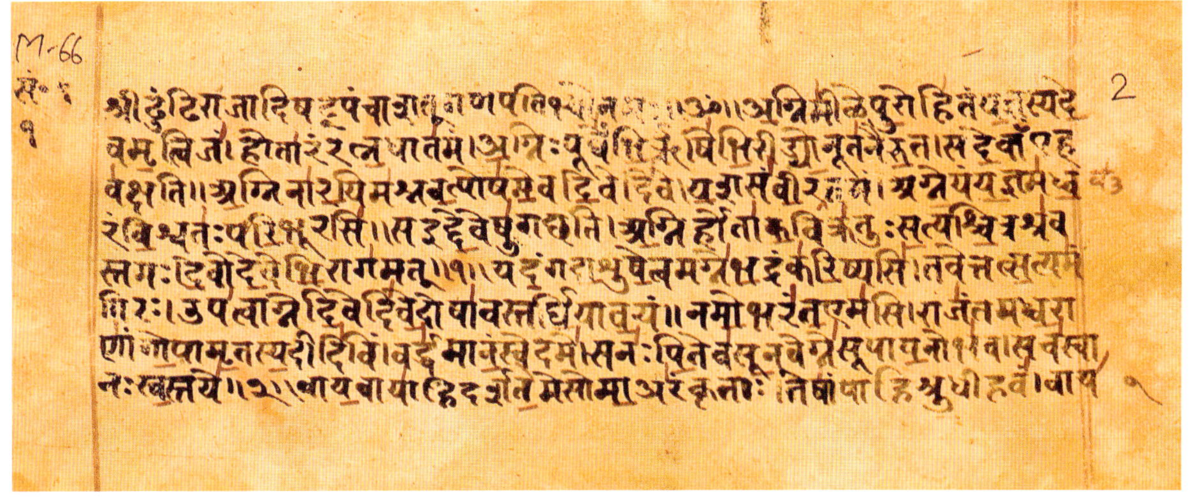

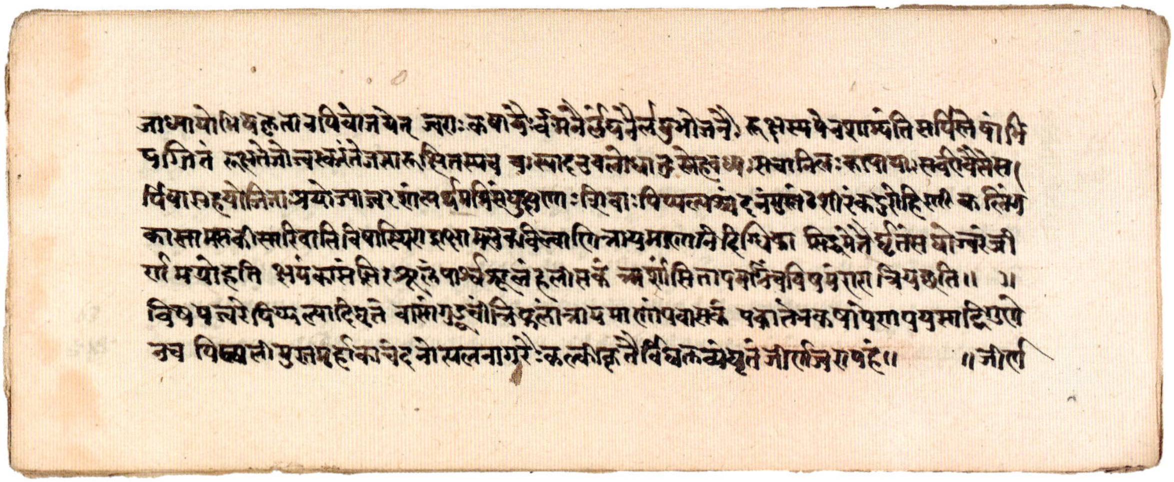

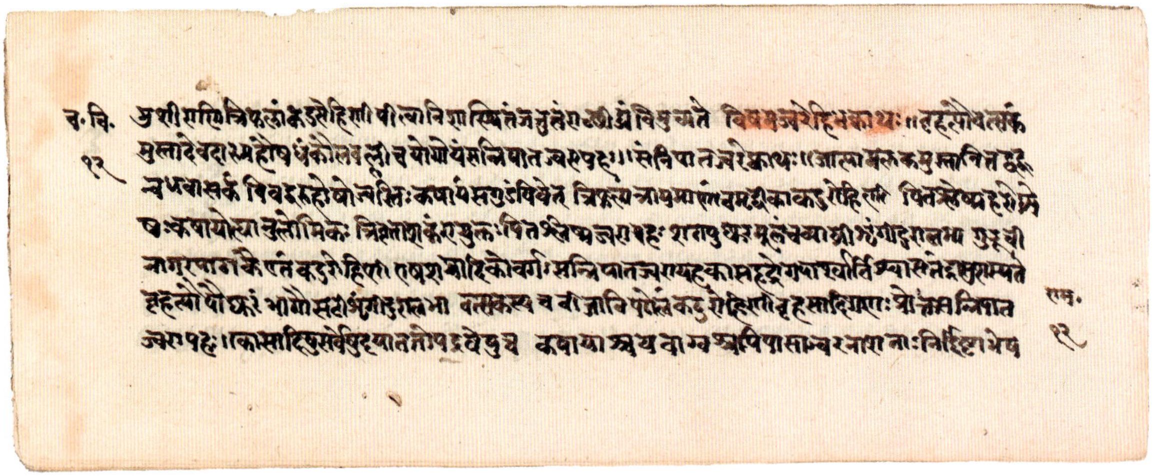

ABOVE

Charaka Samhita,
Charaka and others,
(*c.* 200 BCE–200 CE)

This Sanskrit text is the oldest and most authentic treatise on Ayurveda, India's traditional medicine system.

In the great Indian epic poem, *Ramayana* (Rama's journey, written around 2,300 years ago), most scenes are enacted in forests, and some plants are mentioned in relation to their medicinal use. The most celebrated is *sanjeevani*, which is credited with saving the life of Rama's brother Lakshmana after he is struck by a spear. In the poem, the plant is said to be able to bring people back to life. Today, *Selaginella bryopteris* is considered a possible candidate for the *sanjeevani* plant. It is used in India to treat patients who are unconscious, as well as those suffering with intestinal and urinary problems. Recent studies have found that the plant can provide relief from heat stroke.

Ayurveda, considered to be one of the oldest traditional systems of medicine in the world, has its roots in the *Vedas*. The legendary sage Agnivesha is said to have compiled the medicinal knowledge from the *Vedas*, which was later edited by Charaka (meaning wanderer or wandering physician) and other scholars around 2,200 to 1,800 years ago to form what is now known as the *Charaka Samhita*. It comprises more than 8,400 verses of poetry. Focusing on internal medicine, it presents Ayurveda (which means life knowledge or science) as a comprehensive healthcare system aimed at both preventing and curing disease. The earliest versions of the manuscript that survive today are based on a text completed by Dridhabala around 400 CE.

The *Sushruta Samhita*, produced in *c.* 600 BCE, presents the field of Ayurvedic surgery. The earliest existing version of the *Sushruta Samhita* is a palm-leaf manuscript dated to 878 CE that resides in the Kaiser Library in Kathmandu, Nepal. The *Charaka Samhita* and *Sushruta Samhita* provide detailed descriptions of 620 and 775 plants, respectively. They present information on various aspects of the plants, including when and how they should be collected, how they can be classified (for example, according to their purgative and astringent properties), the parts of plants that are useful, how they can be used therapeutically and their pharmacology. Together these works form the foundation stones of Ayurveda, and are still used today by practitioners of traditional medicine.

BELOW

***Sushruta Samhita*, Sushruta, (600 BCE)**

This text exhibits detailed knowledge of human anatomy, alongside a pharmacopeia of 775 medicinal plants.

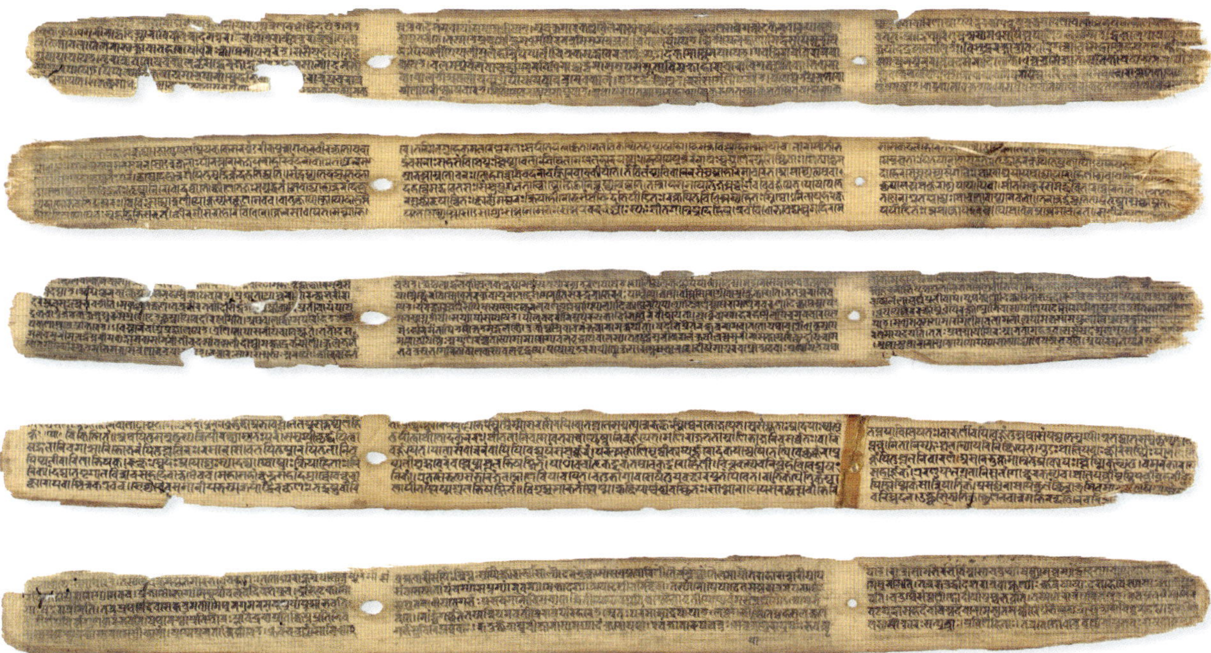

3 Greeks launch the study of botany

As the authors of the *Charaka Samhita* were setting out the principles of Ayurveda, so the foundations of Western science were being laid in Greece. A key text in this regard is the *Hippocratic Corpus*. Referencing the physician Hippocrates, who lived in the fifth century BCE, it is a series of medical treatises that document the changing medical know-how of a collective of itinerant physicians over seven centuries from Hippocrates' day. With no regulation of medicine at this time, the people of ancient Greece would have been at the mercy of a motley crew of bonesetters, midwives, folk practitioners and religious healers for their healthcare. The Hippocratic authors' focus was on rational, rather than supernatural causes of disease, employing high moral standards and healing patients using simple preparations made with locally collected plants. Hippocrates himself is credited with saying: 'Let food be your medicine and medicine be your food'. Nearly half of the documented 3,480 plant uses in the *Corpus* come from forty-four plants, thirty-four of which are also eaten as food. Food plants that feature frequently in remedies include celery, cabbage, barley, lentils, millet, sage, oregano, pomegranate and blackberry.

The *Hippocratic Corpus* was among many intellectual and artistic achievements to emerge from ancient Greece. At the heart of the development of new branches of thinking

BELOW LEFT

***Historia Plantarum*, Theophrastus, (1644)**

This text was written between 357 and 287 BCE, and subsequently much duplicated. This frontispiece is from an illustrated version of the work printed in 1644.

BELOW RIGHT

***Hippocratic Corpus* (1559)**

This collection of around 70 medical works outlined treatments that could be made from locally available plants.

THEOPHRASTI ERESII
DE
HISTORIA PLANTARVM
LIBRI DECEM.
Græcè & Latinè.
In quibus
Textum Græcum variis Lectionibus, emendationibus, hiulcorum supplementis. Latinam GAZÆ versionem nova interpretatione ad margines totum Opus absolutissimis cum Notis tum Commentariis item variorum Plantarum iconibus illustravit
IOANNES BODÆVS a STAPEL.
Medicus Amstelodamensis.
Accesserunt
IVLII CÆSARIS SCALIGERI.
in eosdem Libros Animadversiones
ET
Indice locupletissimo.
AMSTELODAMI
Apud Henricum Laurentium
Anno 1644.

LIVRE
D'HIPPOCRATES,
DE LA GENITVRE DE
l'homme, traduict de Grec & mis en François
par Guillaume Chrestian, medecin ordinaire du
Roy & de messeigneurs ses enfants.
A PARIS, M. D. LIX.
Chez Guillaume Morel, imprimeur du roy.
AVEC PRIVILEGE.
No. 78. Sylvius. 1559.

were the Academy, founded by the philosopher Plato in 387 BCE and considered to be the first institution of higher learning in the Western world, and the peripatetic school at the Lyceum, established by the polymath and philosopher Aristotle around 335 BCE. From these centres and their students came advances in philosophy, physics, astronomy, metaphysics, ethics, chemistry, biology, zoology – and botany. Indeed, the word 'botany' derives from the Greek *botane*, meaning grass or pasture. Two early texts of significance for the development of botany as a discrete field of study are the nine-volume *Historia plantarum* (*Enquiry into Plants*) and the six-book *De causis plantarum* (*Causes of Plants*). Part of a huge body of work spanning natural science, science and human characteristics – much of which is now lost – these were the work of Theophrastus, a student and close friend of Aristotle.

Theophrastus observed how plants looked: collecting, recording and dissecting specimens, and gathering information on their habit and character and the influence of climate or location. He spoke to local experts – farmers, fishermen, shipbuilders, herb-dealers and druggists – and exchanged letters with other intellectuals across the Greek-speaking world. In *Historia plantarum*, he used this information to classify and describe plants, considering trees, plants of different regions, shrubs, herbs and cereals, as well as assessing plants' juices and medicinal applications. In *De causis plantarum*, he turned his attention to plant functionality, considering how plants sprout, flower and fruit. Theophrastus followed the divisions laid out in Aristotle's zoological works, which also frequently mention plants. The text *De plantis* (*On Plants*) was originally attributed to Aristotle but is now known to be the work of the polymath Nicolaus of Damascus, drawing heavily on the material of Aristotle and Theophrastus.

As with the *Hippocratic Corpus* authors, Theophrastus sought to be rational, and negotiate a path between observation, superstition and practical knowledge. He cautioned, for example, that while some of the received wisdom on digging and harvesting roots might be valid, other information could be exaggerated. He noted that the instruction to stand windward when cutting some roots, such as *Thapsia*, and to 'first anoint oneself with oil, for one's body will swell up if one stands the other way', was 'not off the point' as the 'properties of these plants are harmful', taking hold 'it is said, like fire and burn'. Today, the sap of *Thapsia* (a genus also referred to as 'the deadly carrots') is known to be phototoxic, and can cause red welts on skin exposed to sunlight. However, Theophrastus rejected as 'absurd' the idea that cutting a human-like mandrake root must involve circling it with a sword and talking of love while dancing around the plant.

Theophrastus included plants from places outside of Greece, such as Egypt, the Sudan (then Ethiopia) and India, in his works. Greece had been trading with Egypt from before Theophrastus' time, with the former receiving grain, linen and papyrus and the latter gaining Greek wine. As a result, the land and flora of Egypt were reasonably familiar to the Greeks. This shows in Theophrastus' works as he frequently refers to Egyptian plants by name, either in Egyptian or using a Greek moniker. India, by comparison, stood at the eastern end of the known world in Theophrastus' lifetime, and the Greeks knew little about it other than that it had been reported as being highly fertile. Details of its rich vegetation only began to make it back to Greece after Alexander the Great's army reached the Indus around 331 BCE, which coincided with when Theophrastus was working on

TOP

Aristotle,
(384–322 BCE)

This ancient Greek philosopher classified organisms into those that did not move (plants) and those that did move (animals).

ABOVE

Theophrastus,
(*c.* 371–*c.* 287 BCE)

A student of Aristotle, Theophrastus classified plants into trees, shrubs, under-shrubs and grasses.

Historia plantarum. That the Indian flora was a novelty to him is clear from the fact that he either associated Indian plants with Greek ones or left them unnamed.

Reflecting the oral tradition of information dissemination, ancient Greek authors frequently wrote in verse. One notable poet-physician was Nicander of Colophon, who flourished during the second century. Nicander worked in the entourage of Attalus III, the last king of the Greek city state of Pergamum (in modern-day Turkey), who was particularly interested in plants, poisons and antidotes. Nicander's work *Alexipharmaka* covered twenty-two toxic plants, animals and minerals, including the plant poisons of aconite, hemlock and opium. Students used the easy-to-remember verses of *Alexipharmaka* as part of their studies on toxicology until the Renaissance. Other poems written by Nicander are now lost, while some verses from his work *Georgica* were preserved by the later author Athenaeus of Naucratis. These recall Nicander's observations that mulberries appeared before other fruits; that only a few fungi were edible; and that unripe grapes were astringent and 'drew up the lip'. This work likely influenced the Roman orator Publius Vergilius Maro, or Virgil, who lived during 70–19 BCE, when he wrote his poem *Georgics*. Written in Latin in four books, and naming 164 plants, its themes include cereals and weather, the vine and fruit trees, and horticulture.

4 Compiling knowledge from the Greco-Roman world

From 1200 BCE, the Greeks had been dominating the Mediterranean area. But by Virgil's heyday, the Romans were in their ascendancy, conquering areas of Europe formerly under Greek control. As they did so, they adopted many aspects of Greek society – from their gods and goddesses, to their styles of art, literature and architecture, and their approaches to philosophy, religion and military tactics. This period of Greco-Roman culture was to last for nearly 500 years. Two important botanical texts emerged shortly after Rome had

BELOW

Alexander the Great, (356–323 BCE)

Alexander the Great, depicted in this first-century BCE Roman mosaic, enlightened Theophrastus on plants growing far from Greece.

shifted from a republic to an imperial power. The first, *On the Preparation, Properties and Testing of Drugs*, known by its Latin name *De materia medica* and written between 60 CE and 70 CE by Greek physician Pedanius Dioscorides, is today a prime source of information about the medicines used by the Greeks, Romans and other contemporaneous cultures. The second, *Historia naturalis* (*Natural History*), written between 77 CE and 79 CE, showcases knowledge at the time on a range of themes, including plants.

Dioscorides travelled to Greece, Italy, Asia Minor (Anatolia, Turkey) and Provence in modern-day France, possibly as a military doctor. His journeys provided opportunities for him to study diseases, and to gather and identify medicinal plants and other materials. Many of the plants included in *De materia medica* were previously unknown to Greek and Roman physicians; the author added around 100 plants to the 500 or so (medicinal and otherwise) listed by Theophrastus in *Historia plantarum*. Dioscorides believed it was vital to observe plants in their natural habitats and at different life stages, and he criticized contemporary authors who simply discussed earlier knowledge and, in one case, 'obtained his information from erroneous gossip'. In his view, such field studies were important, as the conditions of plants' natural habitats could influence their medicinal potency. He noted, for example, that: 'specific medicinal herbs are stronger or weaker if found on hills and mountains; if exposed to winds; if the position is cool and arid – their strength can rest entirely on such conditions'.

Dioscorides classified the plants he recorded from a purely practical perspective, according to their properties and uses. The five books that make up *De materia medica* cover themes including: aromatic plants; growths providing oily, gummy or resinous products for use in salves and ointments; fleshy fruits; cereals; roots, juices, herbs and seeds for food and medicine; narcotic and poisonous medicinal plants; and vines and wines. Some animal products and metallic ores are also examined. The text describes some 1,000 medicinal remedies, for applications from inducing an abortion and treating intestinal pains, to healing skin and eye diseases and curing toothache. As with today, the adulteration or substitution of medicinal materials was a problem: Dioscorides reports frankincense being adulterated with pine resin and gum, for example.

Myths accompany rational observations in *De materia medica*. Black hellebore had to be dug up with care lest an eagle observed the act – causing death. And, by now, the sword dance required to harvest mandrake in Theophrastus' day had been superseded by the use of dogs to haul the root, screaming, from the ground. The book was influential from the outset, being cited by the Greek physician Galen of Pergamum in around 180 CE in his *De simplicium medicamentorum temperamentis ac facultatibus* (*Mixings and Powers of Simple Drugs – or 'Simples'*), which covered medicine, pharmacy and drugs, and detailed plants used in remedies.

Gaius Plinius Secundus, better known as Pliny the Elder, took an expansive approach when writing his thirty-seven-book encyclopaedic text *Historia naturalis*, including information on cosmology, animals, magic, botany, *materia medica*, mining and minerals. Born in Como, in Cisalpine Gaul (in modern-day Lombardy, Italy), Pliny served in the military and as a governor of Gallia Narbonensis (in modern-day southern France), as well as writing numerous texts, of which only *Historia naturalis* has survived. He was described by his nephew Pliny the Younger, as 'combining a penetrating intellect with

RIGHT

Historia naturalis,
Pliny the Elder,
(1472)

Frontispiece of an early printed version of this encyclopaedic text, which is one of the most substantial ancient works still available to modern readers.

OPPOSITE

Pliny the Elder,
(23–79 CE)

Pliny the Elder was a military commander, author and naturalist during the early years of the Roman Empire.

C·PLINII SECVNDI NOVOCOMENSIS NATVR
ALIS HISTORIÆ LIBER PRIMVS

C·PLINIVS DOMITIANO·IMP·S·

IBROS NATVRALIS HISTORIAE NO-
uitium camœnis quiritium tuorum opus natum
apud me proxima fœtura licentiore epiſtola narrare conſtitui tibi iucundiſſime imperator. Sit.n.
hæc tui præfatio ueriſſima: dum maxio cõſeneſcit
in patre. Nãq; tu ſolebas putare eſſe aliqd meas
nugas: ut obiicere moliar Catullum conterraneũ
meum. Agnoſcis & hoc caſtrẽſe uerbum. Ille enĩ
ut ſcis: permutatis prioribus ſyllabis duriuſculũ
ſe fecit: q̄ uolebat exiſtimari a uernaculis tuis: &
famulis. Simul ut hac mea petulãtia fiat: quod
proxime nõ fieri queſtus es: ĩ alia procaci epiſto-
la noſtra: ut in quædam acta exeãt. Sciantq; omnes: q̄ ex æquo tecum uiuat impium.
Triumphalis & cenſorius tu ſextumque conſul ac tribuniciæ poteſtatis particeps. Et
quod iis nobilius feciſti: dũ illud patri pariter & equeſtri ordini præſtas præfectus præ-
torii eius: omniaq; hæc reipub. Et nobis quidem qualis in caſtrẽſi contubernio? Nec
quicq̄ mutauit in te fortunæ amplitudo in iis: niſi ut prodeſſe tantundem poſſes: &
uelles. Itaq; cum cæteris in ueneratione tui pateant omnia illa: nobis ad colendum te
familiarius audacia ſola ſupereſt. Hanc igitur tibi imputabis: & in noſtra culpa tibi
ignoſces. Perfricui faciem: nec tamen profeci. Quando alia uia occurris ingens. Et
longius etiam ſubmoues ingenii facibus. Fulgurat in nullo unq̄ uerius dicta uis elo-
quentiæ. Tibi tribunitiæ poteſtatis facũdia. Quãto tu ore patris laudes tonas? Quãto fratris amas? Quantus in poetica es? O magna fœcunditas animi. Quẽadmodũ
fratrem quoq; imitareris: excogitaſti. Sed hæc quis poſſet intrepidus æſtimare? ſubi-
turus ingenii tui iudicium: præſertim laceſſitum? Neque enim ſimilis eſt conditio
publicantium: & nominatim tibi dicantium. Tum poſſem dicere: quid iſta legis imperator? Humili uulgo ſcripta ſunt: agricolarum: opificum turbæ: deniq; ſtudiorũ
ocioſis. Quid te iudicem facis? Cum hanc operam condicerem: non eras in hoc albo.
Maiorem te ſciebam: q̄ ut deſcenſurum huc putarem. Præterea eſt quædam publica
etiam eruditorum reiectio. Vtitur illa &. M. Tullius extra omnem ingenii aleam po-
ſitus. Et quod miremur: per aduocatum defẽditur. Hæc doctiſſimum omniũ Perſiũ
legere nolo. Lælium Decimum uolo. Quod ſi hoc Lucillius qui primus cõdidit ſtili
naſum: dicẽdum ſibi putauit. Si Cicero mutuandũ: præſertim cum de repub. ſcribe-
ret: quanto nos cauſatius ab aliquo iudice diffidimus? Sed hæc ego mihi nunc patro-
cinia ademi nuncupatione. Quãplurimũ refert: ſortiatur ne aliquis iudicẽ: an eligat.
Multumque apparatus intereſt apud inuitatum hoſpitem & oblatum. Cum apud
Catonem illum ambitus hoſtem: & repulſis tanquam honoribus ineptis gaudentem:
flagrantibus comitiis pecunias deponerent candidati: hoc ſe facere: pro ĩnocẽtia: quod
in rebus humanis ſummũ eſſet: profitebãt. Inde illa nobilis. M. Ciceronis ſuſpiratio.
O te fœlicem. M. Porti a quo rem improbam petere nemo audet. Cum tribunos ap-
pellaret. L. Scipio Aſiaticus: ĩter quos erat Gracchus: hoc atteſtabãt: uel inimico iudici
ſe approbare poſſe. Adeo ſummum quiſq; cauſæ ſuæ iudicem facit: quẽcunq; eligit:
Vnde prouocatio appellatur. Te quidem in excelſiſſimo humani generis faſtigio po-
ſitum ſumma eloquentia ſumma eruditione præditum religioſe adiri etiam a ſalutã-
tibus ſcio. Et ideo immenſa præter cæteras ſubit cura ut quæ tibi dicãtur: cum digna

amazing powers of concentration', and who 'made extracts of everything he read, and always said there was no book so bad that some good could not be got out of it'. This approach resulted in *Historia naturalis* comprising a mixture of information, some of which was accurately portrayed from direct observations, but large parts of which derived from secondary – often out-of-date – sources.

Pliny cited a large number of such sources, encompassing major authors, lesser authorities and anonymous citizens. He depended heavily on Theophrastus for descriptions of plants, and also drew on the work of Sextius Niger, a Roman writer on pharmacology, for '*materia medica*'. Interested in how plants could be useful to humankind, Pliny wrote about papyrus and the invention of paper; perfumes and their manufacture; vines and viticulture, including the physiological effects of wine (such as when Alexander the Great killed two friends in a drunken stupor); the olive tree and olive oil production; and cereals, bread and baking. For these sections, he consulted the works of Virgil, Marcus Terentius Varro (who wrote on many topics, including *Res rusticae/On Agriculture)*, Marcus Porcius Cato (Cato the Elder; author of *De re rustica/On Agriculture*, covering farming, rituals and recipes) and Marcus Columella (author of *Res rustica/On Agriculture* and *De arboribus/On Trees*). Pliny's own observations include his examination of almost all the plants within the garden of the expert Roman horticulturist Antonius Castor.

Pliny died in 79 CE, his unstoppable curiosity having led him to linger too long while watching Mount Vesuvius erupting, but some three centuries later an unknown author or authors compiled parts of his work into a text that now is referred to as pseudo-Apuleius' *Herbarius*. Although credited to Apuleius, a second-century writer on science and philosophy from Roman-ruled Algeria, this was likely done to lend authority to the work, hence the 'pseudo' prefix. The herbal devotes a chapter to each of 131 plants, providing synonyms, growing locations and recipes for treating various ailments. There is evidence that the herbal may have drawn on information in either *Historia naturalis* or *Medicina Plinii* (an early fourth-century compilation of herbal remedies based on *Historia naturalis*). The text recommends giving the juice of the leaves of victoriola (Dutch Butcher's Broom *[Ruscus hypophyllum]*) with honey to drink on an empty stomach for destroying phlegm; relieving inflammation and pain in the eyes by smearing them with the juice of priapiscus (ground orchid bulbs); and applying the juice of artemisia with oil of roses to dispel fevers.

The preface of the pseudo-Apuleius' *Herbarius* suggests it was intended as a self-help manual to avoid the sick having to rely on untrustworthy physicians. Reflecting Pliny's distrust of medical practitioners, it says:

> **From several public documents, I have related a few virtues of plants and bodily remedies, in truth, because of the verbose stupidity of the medical profession, what we say are the "huckstering rather than the cures" of physicians, and also that these men, in general zealous with their ignorance and incompetence, may truly be called greedy, who even claim a fee from their dead patients. What do they do? Nothing. Indeed, they wait for opportunity and they make money as long as they draw out the length of their patient's treatment, because, in my opinion, they are more dangerous than the diseases themselves.**

5 Medieval monasteries duplicate earlier learning

By now, in the fourth century CE, the Western Roman Empire (comprising today's Spain, Portugal, France, Belgium, the Netherlands, Switzerland and much of Britain) was in decline, and Christianity was in its ascendancy. In 476 CE, the Western Roman Empire fell to Germanic conquerors, ushering in the Medieval period. No longer controlled by a central government, this vast land area came to be run by various vying groups. The advances in botanical understanding made by the ancient Greeks had led Dioscorides to write of 'the botanical tradition', referring to the study and application of medicinal plants. Now the practice of physically observing plants was lost, as curiosity in the natural world waned, and the Catholic church came to dominate. For the next thousand years, the majority of botanical texts were concerned solely with practically using plants for healing, and were grounded in the work of the earlier Greek and Roman authors.

Dioscorides' *De materia medica* is among the most duplicated texts from antiquity. The finest surviving copy, known as the *Codex Vindobonensis*, is in the original Greek language, and has 400 magnificent full-page illustrations of plants. It was made around 512 CE in Constantinople (modern-day Istanbul, Turkey), which at the time was the capital of the Eastern Roman – or Byzantine – Empire. The *Codex* was gifted to Patricia Juliana Anicia, daughter of Flavius Anicius Olybrius – who had briefly been emperor of the Western Roman Empire in 472 CE. The manuscript re-emerged in 1406, when it was rebound. It came into the hands of the Turks after Constantinople fell to the Ottoman Empire in 1453, before being given to the Holy Roman Emperor in Vienna in the sixteenth century. Eleven of the images in the Codex appear to be copied from illustrations made by pharmacologist and physician Krateuas, who lived in the late second and early first century BCE.

BELOW

***De materia medica,* Dioscorides, (60–70 CE); *Codex Vindobonensis* version (c.512)**

Heuresis, the personification of discovery, presents Dioscorides with a mandrake that has been plucked from the ground using a dog, as was the custom.

At that time, plants had long featured in architectural decoration but the idea of illustrating herbals as an aid to identifying plants was relatively new. Pliny had been dismissive of attempts by Krateuas and his fellow artists Dionysius and Metrodorus to do so, writing: 'Not only is a picture misleading when the colours are so many, particularly as the aim is to copy Nature, but besides this, much imperfection arises from the manifold hazards in the accuracy of copyists.' He made a valid point; the *Codex* images were almost certainly copied, and, while they look naturalistic, they do not entirely correspond with nature. Such inaccuracy was to remain an issue in botanical illustration for the next 1,500 years. Today, the *Codex* resides in Vienna's Austrian National Library. As one of the earliest surviving, relatively complete and illustrated codices on any medical or scientific subject, it provides unique insights into early plant science and art.

Soon, it had been fully translated from Greek into five Latin books. At around this time, the *Ex herbis femininis* (*Medicines Made From Female Herbs*) was also produced, detailing seventy-one plants and drawing heavily on one of the Latin *De materia medica* translations, and also possibly the pseudo-Apuleius' *Herbarius*. Its name is rather mysterious; although it was common at the time to refer to plants as 'male' and 'female', the text does not focus on female plants.

LEFT

De materia medica,
Dioscorides, (60–70 CE);
(1460 version)

Human-like mandrakes were believed to let out a deadly scream on being uprooted, hence the use of dogs to pull them out.

RIGHT AND OVERLEAF

De materia medica, **Dioscorides, (60–70 CE);** ***Codex Vindobonensis*** **version (c.512)**

Some plants are clearly identifiable from the images. This page shows nettles (left) and asphodel (right); overleaf depicts a rose (left) and shepherd's purse (right).

ΑCΦΟ ΔΕΛΟC

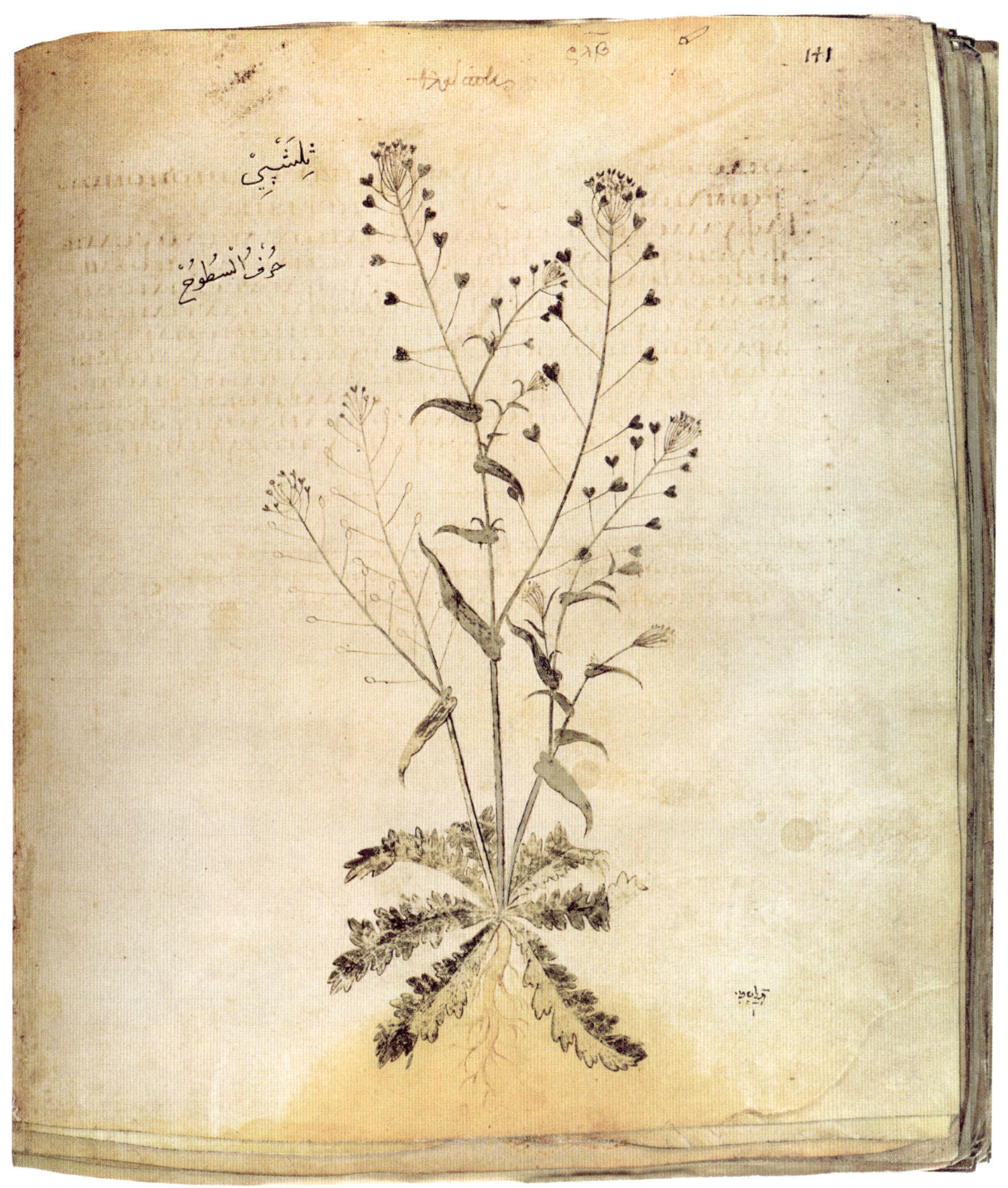

THIS PAGE

***Herbarius,* Pseudo Apuleius, (6th century)**

Arnoglossa (left), which now goes by the Latin name of *Plantago asiatica*, and Dragontea (right), depicted in berry, which is today known as the dragon arum (*Dracunculus vulgaris*).

It is possible that it was conceived as a text on herbs for treating ailments specific to women, but that the copyist misunderstood this meaning. Scholars believe that *Ex herbis femininis* was more popular than the Latin *De materia medica* in the subsequent centuries, because it was shorter, provided more information on the flora of southern Europe and gave more detailed guidance for medicinal use.

During the early Medieval period (800–1000 CE), herbal medicine became entwined with Christianity, with monks being major sources of medical knowledge across Europe. These religious scholars were the prime translators and copyists of ancient texts, but they also updated them with notes informed by their own experiences of using herbs for treating the sick. The Bible – the Old Testament of which had been compiled between 1200 and 165 BCE and the New Testament in the first century CE – could also have been a source of information about medicinal plants at this time. Around five of the forty-five plants mentioned in the text are directly referred to as being of medicinal use. For example, mandrake is referenced in the book of Genesis with regards to enhancing fertility, and hyssop is mentioned in Leviticus in the context of cleansing and purification. Christianity also propagated the view that God was a 'divine physician' who could punish wrongdoing with sickness and cure those who repented. However, this was a reflection of the rivalry with the pagan Asclepius cult rather than a response to the healing work of Jesus Christ.

BELOW

Hebrew Bible
(*c.* 1300)

The **Bible** provides scholars with information on which plants were in use in the Middle East from 1200 BCE.

One text written in England, *Bald's Leechbook* (*Book of Medical Prescriptions*), sheds light on the extent of Medieval medical knowledge. Written in Old English (at a time when Latin was the language of scholarship in much of western Europe) the three-part text contains old English charms and remedies that combine Mediterranean medical know-how from the third to the ninth centuries with local practices. While the text highlights how spiritual and magical elements were an integral part of healing during this period, it also contains potentially effective medical remedies. For example, in 2015, a study by the University of Nottingham, UK, showed that the *Bald*'s eye salve (a treatment for healing a stye, made with garlic, wine and bovine bile), could combat the bacteria MRSA. Interestingly, notations on the manuscript show that the chapter on diseases of the eye and the dimming of eyesight, in particular, remained of interest to readers into the late Medieval period (1300–1500).

6 Knowledge of plants expands in China

While the centuries following the fall of the Western Roman Empire were a time of relative stagnation in botanical knowledge and exploration in Europe, interest in plants and their uses was increasing in Asia. Reputedly, the oldest work on tropical and subtropical botany is the *Nan-fang ts'ao-mu chuang* (*Plants of the Southern Regions*), which was written in 304 CE by Chi Han (263–307 CE). Covering southern China and parts of Vietnam, the text discusses eighty plants, classifying them as herbs, trees, fruits and bamboos. At the time the text was written, civil wars were widespread in the north of China, but the south enjoyed relative peace and prosperity, and was connected to the West through maritime routes. The book includes a number of plants originating in India or western Asia, such as black pepper and jasmine. It also provides interesting ethnological insights, including how Chinese spinach was planted on floating rafts made of reeds, and that fine cloth was fashioned from banana fibre.

BELOW

Bencao mengquan
Chinese woodcut, (1573–1620)

The mythical Chinese emperor Shen Nong is said to have had the head of a bull and the body of a man.

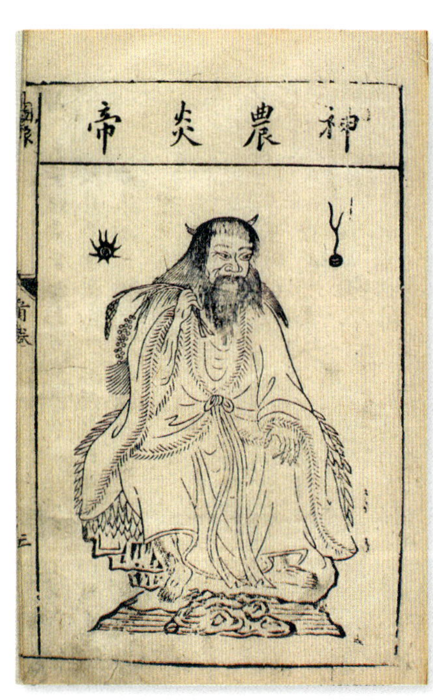

Modern understanding of events in China in the centuries after Chi Han's time was enhanced when tens of thousands of manuscripts, dating from the fifth to the eleventh centuries, were discovered in the early twentieth century in a hidden cave at the Buddhist complex of Mogao at Dunhuang in Gansu Province. Encompassing works of historic, religious, scientific, artistic and linguistic importance, they include the preface and fragments of a text, known as the *Ben Cao Jing Ji Zhu* (*Collection of Commentaries on the Classic Materia Medica*), written in 500 CE by the Chinese alchemist and pharmacologist Tao Hongjing (456–536 CE). In it, Tao Hongjing preserved and annotated an earlier text *Shen Nong Ben Cao Jing* (*The Divine Husbandman's Classic of Materia Medica*) which had been written during the Eastern Han Dynasty between 25 CE and 220 CE.

The earlier text had summarized the history of herbal medicine in China, initially passed down orally, up to the Han Dynasty. It formally documented 365 medicinal herbs that, at the time, were in

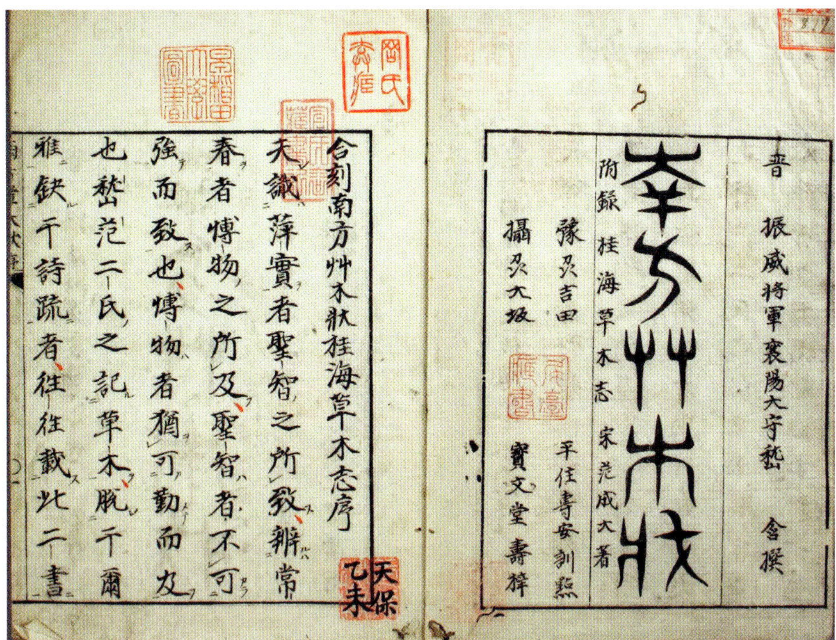
晉 振威將軍襄陽太守嵇 含撰

南方艸木狀

附錄 桂海草木志 宋范成大著

豫 及吉田 平住專安訓點

攝 及大坂 寶文堂壽梓

合刻南方艸木狀桂海草木志序
天識萍實者聖智之所致辨常
春者博物之所及聖智者不可
強而致也博物者猶可勤而及
也嵇范二氏之記草木脫于爾
雅缺于詩疏者往往載此二書

天保乙未

LEFT

Nan-fang ts'ao-mu chuang (Plants of the Southern Regions), **Chi Han,** (1726)

This work recorded plants including jasmine, yam and water lily.

common medicinal use, grading them in three ranks according to their curative activity and toxicity. The text was compiled by a number of authors over a period of time but credited to Shen Nong, the mythical 'Divine Husbandman' and folklore deity. Shen Nong is said to have established agriculture in China some 5,000 years ago, becoming fascinated by the curative properties of plants in the process. Tasting herbs to ascertain their effect, he selected those with medicinal activity and described their properties. Among them was cannabis – the earliest recorded medical use of the plant – which Shen Nong considered to be a magic mood and memory booster. Some versions of the myth about Shen Nong state that he died after ingesting a particularly toxic plant.

Tao Hongjing's update of the text doubled the number of medicinal substances to 730, drawing on a corpus of writings he described as 'separate records of eminent physicians'. He categorized these substances under the seven headings of: jades/stones, herbs, trees, insects/beasts, fruits/vegetables, crops and those with names but without applications, retaining the earlier rankings as sub-headings. He also advanced on methods of treatment. *Shen Nong Ben Cao Jing* had promoted using cooling drugs for patients suffering with fever, and drugs with hot properties for combatting cold symptoms. This continues in Traditional Chinese Medicine (TCM) today; a typical cooling medicine used to treat hot symptoms is zhimu, the dried rhizome of *Anemarrhena asphodeloides* (in the asparagus family), while an example of a heating drug used to treat chills is fuzi, the processed lateral root of wolf's bane (*Aconitum carmichaelii*). However, Tao Hongjing felt that Shen Nong's approach to addressing patients' symptoms should be broader. He considered that applying the appropriate dose of the correct kind of medicine called for a more holistic assessment encompassing the nature of the patient's sickness, as well as their gender, age, overall physical condition and the climatic conditions of their dwelling.

Tao Hongjing lived in southern China under the Liang Dynasty at a time when the north was ruled by the Northern Wei Dynasty. This prevented him from including information about substances used as drugs in the northern parts of China. Nonetheless he initiated a 'main tradition' of comprehensive *Ben Cao* works that continued under different authors for another six centuries. In each, the original classic text was expanded on, with new data presented on the origin, medicinal effects and required processing of the showcased substances. Often these works incorporated information that was outside of the main tradition, such as knowledge on regional plants or those that could support populations in times of famine. The first to follow Tao Hongjing was the *Xin Xiu Ben Cao* (*Newly Revised Materia Medica*), which was commissioned by the government of the Tang Dynasty and published in 659 CE. This recorded 850 medicinals from both the north and the south, and is considered to be China's first national pharmacopoeia.

BELOW

Tang ye Ben Cao (*Materia medica of decoctions*), Wang Haogu, (1298–1308)

This text described how decoctions, powders and pills could be applied for a fast or slow effect.

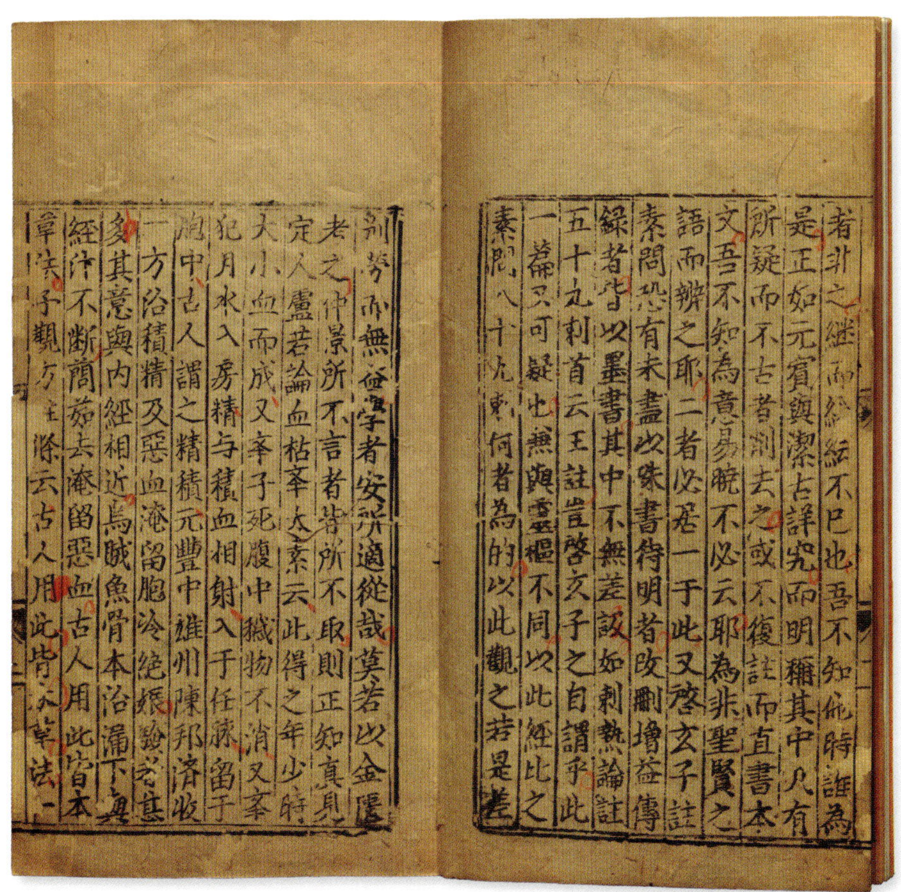

者非之繼而紛紛不已也吾不知何時誰為
是正如元賓與潔古詳究而明稱其中凡有
所疑而不古者削去之或不復註而直書本
文吾不知為意易曉不必云耶為非聖賢之
語而辨之耶二者必居一于此又啓玄子註
素問恐有未盡以朱書待明者改刪增益傳
錄者皆以墨書其中不無差誤如刺熱論註
五十九刺首云王註豈啓玄子之自謂乎此
一篇又可疑也無與靈樞不同以此經比之
素問八十九刺何者為的以此觀之若是差

剐勞而無益學者安所適從哉莫若以金匱
老之仲景所不言者皆所不取則正知真見
定人盧若論血枯乎太素云此得之年少時
大小血而成又卒子死腹中械物不消又卒
犯月水入房精与積血相射入于任脉留于
胞中古人謂之精積元豐中雄州陳邦濟收
一方治積精及惡血滲留胞冷絶娠殊驗甚
多其意與内經相近烏賊魚骨本治漏下與
經汁不斷蔺茹去滲留惡血古人用此皆本
章候子觀方注滌云古人用此皆於草法一

This was superseded by *Zheng Lei Ben Cao* (*Materia Medica Arranged According to Pattern*), a compendium of 1,476 medicinals compiled by Tang Shenwei (*c.* 1056–93) and reissued in 1108 under the Song Dynasty. Having survived intact, it is an important source of knowledge on medicinal practices up to this time. By now, though, the approach of simply adding new material to old was becoming unwieldy and soon died out. It was replaced by a new genre of *materia medica* texts, which focused on individual aspects of herbal medicine rather than attempting to cover everything. An example is Wang Haogu's thirteenth-century *Tang Ye Ben Cao* (*Materia Medica of Decoctions*). Around this time, pharmaceutical approaches to healing were united with the practice of needling, or acupuncture, which for the previous 1,000 years had been an isolated part of Chinese medicine.

7 India recognizes the discipline of plant science

In India, as in China, the Medieval period was a time of major advances – in mathematics, astronomy, chemistry, metallurgy, botany and medicine. The term *Vrksāyurvēda* emerged from the ancient literature to mean the science of plant life and plant care. A text bearing this title, written between the first century BCE and fourth century CE, is associated with the great Hindu sage Parasara. It was not written by him; rather it is voiced by him as if he were teaching a student. Unlike remedy-focused European texts of the time, it comprises six parts, which cover the origin of life, ecology, the distribution of forests, the shape, classification and names of plants, as well as their tissue structure and how they function. It exhibits a good understanding of how plants relate to the land, describing *Jāngala* terrain as desert-like with little vegetation and limited water, and *Anupā* as green and grassy with clay soil, rivers, streams, lakes, shorelines, large trees and dense forests.

In the eleventh century another work with the title *Vrksāyurvēda* was produced, by an author known as Surupala. In this text, *Vrksāyurvēda* is explained as following the same principles for plants as Ayurveda does for humans; that is, to preserve or restore the correct balance of three *doshas*: *vata* (air and space); *pitta* (fire); and *kapha* (water). Surupala considered that practitioners should use an Ayurvedic lens to view a plant's condition, and promoted taking a holistic approach to crop management. He noted the importance of appropriate soil, good seed, treating seeds before sowing them, intercropping, timely weeding, optimal water use and using herbal products or dead animals wastes to protect from disorders. He instructed, for example, that: 'Trees that are dried up due to heat caused by fire are cured when a mixture of sugar, sesame and milk is used for watering and anointing them, and when they are smeared all around by mud from the bottom of a lotus plant.'

8 A golden age of Islamic botany

Between the eighth and thirteenth centuries, science flourished within the Islamic world. Islam began in 610 CE, following the first revelation to the forty-year-old prophet Mohammad. Over the following twenty-three years, the Quran was verbally revealed, with the resulting text including some thirty significant plants and providing guidance on conserving nature, such as to only cut down trees in unavoidable situations. For example,

ABOVE
Ibn al-Baytar (1197–1248)

Medieval botanist al-Baytar is commemorated by this statue in his birthplace of Benalmádena, near Malaga, Spain.

on the subject of felling, it says: 'Whether ye cut down/(O ye Muslims!)/The tender palm-trees, or ye left them standing/On their routes, it was/By leave of God . . .'
By 622, the Prophet had established the first Islamic state; this was extended under the Umayyad Caliphate (Muslim government) into parts of Africa, Asia and Europe. In 750, the Umayyad Caliphate was overthrown and replaced by the Abbasid Caliphate, which founded the city of Baghdad in 762. Under the leadership of Harun al-Rashid (r. 786–809), who had led campaigns against the Byzantine Empire before becoming caliph, the city became a global centre of knowledge, culture and trade. Understanding of plants expanded greatly during this period, due to ancient texts being translated and new books emerging.

Under the Umayyad Caliphate, the physician Masarjawayh had translated part of Galen's *Simples* book into Arabic. Caliphs of the Abbasid period sent many scholars to foreign nations to bring back more manuscripts. According to one historian: 'in quest of knowledge men travelled over three continents and returned home, like bees laden with honey, to impart the precious stores which they had accumulated to crowds of eager disciples'. The scientific works they had gathered from Greece, Egypt, the Roman Empire, India and Persia (modern-day Iran) were then translated into Arabic, saving many for posterity. Hunayn bin Ishaq al Ibadi (809–873) and his disciple Staphen Ibn Basilos translated Dioscorides' *De materia medica* into Arabic and Syriac, after which it formed the basis for Arab pharmacopeias; the money-changer-turned-translator Thabit bin Qurrah tackled Nicolaus of Damascus' *De plantis*; and Hunayn took on several of Galen's treatises on medicine and philosophy, as well as Pliny's *Historia naturalis*. When Theophrastus' *Historia plantarum* and *De causis plantarum* were translated, the Arabs noted how ahead of his time the Greek had been.

Muslim scholars also wrote books of their own, adding to the knowledge of plants across the Islamic Empire at this time. Among them are Abu Hanifah Dinawari (d. 895), who wrote the *Kitab al-Nabat* (*Book of Plants*) in the ninth century, drawing on oral and literary sources, including Bedouin poetry. The book provides detailed descriptions of some 1,120 plants – divided into cultivated food plants, fruit-bearing plants and wild plants – mostly covering those indigenous to Arabia (the Arabian Peninsula) but also some brought in from elsewhere. Dinawari also discussed the sky, clouds, moon, sun, rain and winds, along with natural water sources and soils. He refers to plants being male and female, writing: 'Date-palm is so near to animals and even to human beings that sometimes there are love affairs: sometimes a female date palm gets amorous of a certain male date-palm. The female tree even physically inclines towards its beloved'.

Parts of modern-day Spain's Iberian Peninsula were under Muslim control between 711 and 1492. From Granada, Ibn al-Baytar (1197-1248) is considered to be one of the greatest botanists of the Medieval Muslim world. Having studied in Seville, he spent twenty-four years on plant-collecting missions, visiting North Africa, Greece, Asia Minor, Persia, Iraq, Egypt and Syria, and living between Egypt and Syria. In the thirteenth century, he compiled the work *Kitab al-Jami' li Mufradat al-Adwiyah w-al-Aghdhiyah* (*The Book of Medicinal and Nutritional Terms*), an encyclopaedic work summarizing Arab knowledge on botany. Drawing on works by Abu Bakr al-Razi (*c.* 865-925; a physician who promoted opium as an anaesthetic and analgesic), Ibn Sina (980-1037; philosopher, physician and author of *The Canon of Medicine*), Dioscorides, Galen and others, it

LEFT

Oppenheim Quran (14th or 15th century)

The Quran mentions plants that were useful in the 7th century, such as date, garlic, grape and olive.

BELOW

Ibn Sina (980–1037)

The 11th-century Persian physician Ibn Sina was among inspirations for Ibn al-Baytar's 13th-century botanical opus.

RIGHT

***De materia medica*, Dioscorides, (60–70 CE; 13th-century translation)**

A physician prepares an elixir in this Arabic translation of the enduring 1st-century work.

LEFT

***Kitab al-Jami' li Mufradat al-Adwiyah w-al-Aghdhiyah* (*The Book of Medicinal and Nutritional Terms*),**
Ibn al-Baytar,
(13th century)

The author of this work was one of the greatest botanists of the Medieval Muslim world.

RIGHT

***Kitāb fī al-adwiya al-mufrada* (*Book of simple drugs*), Abu Jafar Ahmad ibn Mohammad al-Ghafiqi, (1256)**

This text combined the work of earlier authors, including Dioscorides, Ibn Sina and Ibn al-Baytar.

provided an alphabetic guide to 1,400 medicaments. Ibn al-Baytar included 200 previously unknown plants and used Persian, Berber and Latin names, which encouraged the transfer of knowledge and helped to clarify herb properties and appropriate uses. A further important *materia medica* from this time was Abu Rayhan Mohammad bin Ahmad al-Biruni's *Kitab al-Saydanah fi al-Tibb* (*Book of Pharmacy in Medicine*), which incorporated medicines of India.

9 Arab knowledge revitalizes botanical learning in Europe

From the early eleventh century, a series of religious wars were fought, as European Christians sought to wrest biblical lands from Islamic rule. After the West won the First Crusade, between 1096 and 1099, the inventions and scholarship of the Arabic East, including their promulgation of ancient Greek texts through translation, filtered through to the West and helped stimulate new thinking in European intellectuals of the day. In modern-day Germany, the Benedictine abbess, mystic and visionary, Hildegard von Bingen (1098–1179), promoted the spiritual concept of *viriditas*, or 'greenness', in which the divine was expressed in nature. This was counter to previous views that associated the natural world with Satan. Considered by some to be the first woman to have written about plants, von Bingen drew on her experiences tending monastery gardens and healing the sick to write *Physica* (*Medicine*), describing 500 herbs, plants, animals, precious stones and metals, and providing 2,000 remedies.

BELOW

Hildegard von Bingen (1098–1179)

Hildegard von Bingen was a polymath and an early female author on the many medicinal uses of plants.

In the thirteenth century, the theologian and scientist Albertus Magnus, known as Albert the Great (*c.* 1200–80), argued that theology and philosophy were two separate intellectual endeavours, and sought to train Dominican friars in both. His text *De vegetabilis et plantis* (*On Vegetables and Plants*), primarily a commentary on the text *De plantis* of Nicolaus of Damascus, also incorporated some accurate observations on plants and information on their structure and anatomy. It demonstrated that Albert's interest in herbs went beyond learning about practical uses of them. He also commented on the teachings of Muslim academics, including Ibn Sina.

Certain particular centres helped to disseminate the new thinking that was emerging at this time, among them the Mediterranean island of Sicily, and Salerno, in modern-day Italy. In Salerno, an early medical school had become established. The school trained physicians using resources available at the nearby Benedictine monastery of Monte Cassino. After Sicily and southern Italy came under Norman control from 1130, the monastery's handful of classical manuscripts of the Hippocratic authors and Galen's work was greatly expanded by Latin translations of Arabic texts (some of which also derived from Greek sources), by the monk Constantinus Africanus (d. 1089). Thereafter, teaching at the school combined the practical approach of local Salernitan physicians, founded in ancient Greek traditions, with Arab attention to detail and respect for authority.

Many of the school's faculty also wrote treatises on medicine, which helped to build Salerno's reputation. A key text from the twelfth or thirteenth centuries was *Regimen sanitatis Salernitanum* (*The Book of Health*), a poem commonly attributed to Giovanni da Milano, which included information on diet, hygiene, *materia medica*, bodily structure and functions, and potential treatments. For example, it advises: 'Who would not be Sea-sick when seas do rage, Sage-water drinke with wine before he goes.' Another from the middle of the twelfth century, *Practica secundum Trotam* (*Practical Medicine According to Trota*), was written by a female physician known as Trota of Salerno, who also contributed to texts on female medical issues. These works defied the church by advocating the use of opiates from plants to dull the pain of labour. And Matthaeus Platearius, possibly a son of Trota, produced *Circa Instans* or *Liber de simplici medicina* (*The Book of Simple Medicines*) in the late twelfth century, based on Dioscorides' work.

As Salerno declined as a power in disseminating new medical thought, Toledo, in Spain, rose in importance. In the thirteenth century, the city became a powerhouse of translation, where scholars transcribed texts from Arabic or Hebrew into Castilian, and from Castilian into Latin. These texts fed the minds of students at new centres of learning, such as early universities in Bologna, Paris, Oxford, Cambridge, Salamanca, Montpellier and Padua. The herbals produced thereafter incorporated information from diverse sources, and some included beautiful illustrations. The *Carrara Herbal*, a late fourteenth-century Italian translation of a medical treatise by the Arab physician Serapion the Younger, exhibited highly realistic renditions of plants. The time had come for scholars to shrug off their ingrained habits of repeating old texts in favour of acquiring new knowledge about plants, and to use drawings and paintings from nature to illustrate their findings. Waiting in the wings was a new technology that would help them in this mission, one which would change the world of manuscripts and books forever.

BELOW

Trota of Salerno (12th–13th century)

Trota of Salerno was a female physician and writer who lived in southern Italy in the 12th century.

LEFT

Carrara Herbal
(*c.* 1390–1400)

Based on earlier Latin and Arabic works, the Carrara Herbal presented medicinal 'simples' in the Paduan dialect (this page and following two spreads).

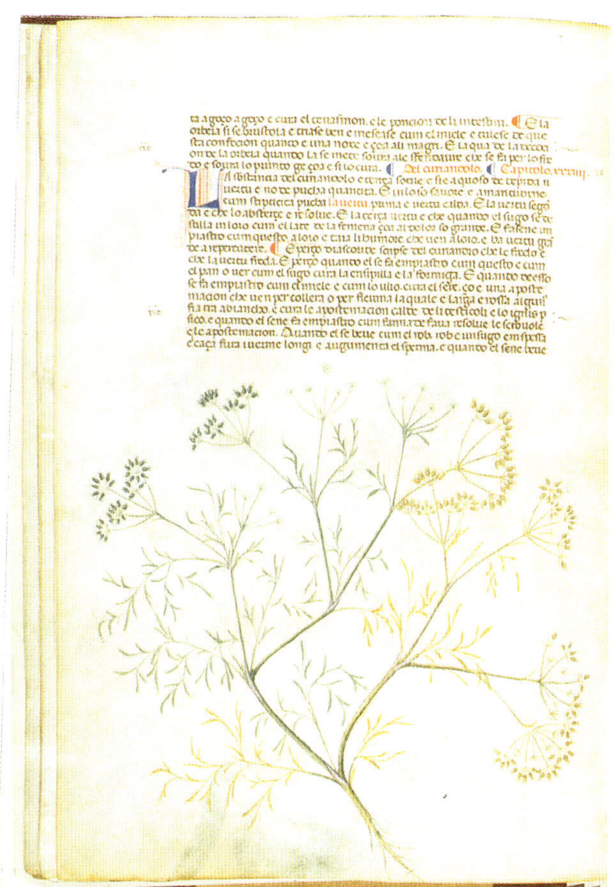

De la camomilla. Capitolo. xcii.

2

THE COMING OF PRINT
(1450–1600)

The study of plants was changed completely by printed books, a global shift in the way knowledge was owned, distributed and used. All the elements of the movable type process had been pioneered in China and Korea from the 800s onwards, but in 1454 in Mainz, Johannes Gutenberg (*c.* 1400–68) put together a production system of printing that was to spread across the world. Instead of each page of a book needing to be cast or woodcut or written as one piece, now words and letters could be printed, disassembled, sorted and reused at dizzying speed.

Even a luxurious printed book was a fraction of the price of a manuscript. By the year 1500, a few decades after printing had begun, some estimated ten million books had been printed in Europe, 40,000 different titles, for a population of no more than 100 million. Printing was concentrated at first in a few European cities: Mainz, Antwerp, Venice, Paris and London. Such cities frequently offered monopolies on printing to favoured individuals, a keen reminder of the medieval guild systems of protection for skilled workers. Strict rules were also laid down by religious authorities, such as the Pope, to prevent printing of religious materials without Church authorization. When God's creation was considered to comprise all things, the Church might scrutinize even non-religious works, as theologian and botanist Otto Brunfels (1488–1534) would later learn when all his works were banned by the Vatican, including those that were solely botanical, due to his Lutheran beliefs.

Few presses existed outside major cities even by 1600. But the skills and technology spread in intriguing ways: the first presses reached Morocco in Africa in 1516 under the aegis of Portuguese Jews, who had learnt printing in Lisbon but fled when their faith came under persecution. Presses were loaded onto Portuguese ships bound for India, Macao and, by 1590, Nagasaki. As the celebrated French historian Fernand Braudel (1902–95) wrote: 'If the invention had indeed originated in China, the process had gone full circle.'

The movement of printing also echoes and shadows the routes of voyages of contact and investigation made by Europeans in the fifteenth century and onwards. Journeys made by travellers ranging from crusaders to imperial emissaries widened the zones of contact between Europe and Asia. Some of this was motivated by an interest in plant material – having long seen and traded in sought-after spices that arrived on overland routes, European nations had sought new maritime routes to their sources. Longer exploratory voyages, initially led by Portugal and Spain reached Africa and the Americas, where quickly issues of colonial dominance came to the fore.

Trade routes were an intense preoccupation for rivalrous European rulers, hoping to outdo each other; many were motivated by the promise of untold riches from a new world, particularly spices. Travellers returned with botanical gold in the form of new crops and medicinal plants; for example the Francisco Hernández expedition of 1571–77, brought back for the first time tomatoes and cacao (see page 115). Back in Europe, printers and authors vied to be the first to present these new discoveries. Printing was a way for ambitious nations and individuals to mark their stamp on the world, with priority in publishing becoming a way of staking a claim to knowledge, territory and the earth itself.

PLINIVS secundus nouocomensis equestribus militiis industrię functus: procu-
rationes quoq; splendidissimas atq; continuas summa integritate administrauit. Et
tamen liberalibus studiis tantam operam dedit: ut non temere qs plura inotio scripserit.
Itaq; bella omnia quę undiq; cum romanis gesta sunt .xxxvii. uoluibus comprębendit. Itē
naturalis bistorię .xxxvii. libros absoluit. Periit gadis campanię. Nam cum misenēsi classi
pręesset & flagrante Vesuuo: ad explorandas propius causas liburnicas pręstendisset: neq;
aduersantibus uentis remeare posset: ui pulueris ac fauillę oppressus est: uel ut quidam
existimant a seruo suo occisus: quē deficiens ęstu ut necem sibi maturaret orauerat bic in
bis libris .xx. milia reꝝ dignaꝝ ex lectione uoluminū circiter duum milium cōplexus est.
Primus aūt liber quasi index .xxxvi. libroꝝ sequentium consumationem totius operis &
species continet tituloꝝ.

CAIVS PLINIVS SECVNDVS DOMICIANO I-
PERATORI SALVTEM PLVRIMAM DICIT:

LIBROS NATVRALIS HISTORIAE
nouitiū camęnis qritiū tuoꝝ opus natū apud me
proxima fętura licentiore epistola narrare cōstitui
tibi iocūdissime imperator. Sit enim bęc tui prę-
fatio uerissima: dum maxime consenescit i patre.
Namq; tu solebas putare esse aliqd meas nugas:
ut obicere moliar Catullum conterraneū meum
agnoscis & boc castrēse uerbum: ille enim ut scis
pmutatis prioribus syllabis duriusculum se fecit
q̃ uolebat existimari a uernaculis tuis & famulis.
Simul ut bac mea petulantia fiat ꝙ proxime non
fieri questuses in alia procaci epistola nostra ut in
quędam acta exeant. Sciantq; omnes q̃m exęquo
tecum uiuat iperium triumphalis & censorius tu
sexiesq; cōsul ac tribunitię potestatis particeps: et
ꝙ iis nobilius fecisti: dū illud patri pariter & eques-
tri ordini pręstas pręfectus prętorii eius omniaq;
bęc rei publicę: et nobis quidē qualis incastrēsi contubernio. Nec qcq̃ mutauit ite fortunę
amplitudo in bis: nisi ut prodesse tantundē posses ut uelles. Itaq; cū cęteris iueneratione
tui pateant omnia illa: nobis adcolēdum te familiarius audatia sola supest. Hanc igit tibi
imputabis: et in nostra culpa tibi ignosces. Perfricui faciē: nec tamē profeci quoniam alia
uia occurris: igens & longius etiam summoues igētibus fascibus fulgorat in nullo unq̃
uerius dicta uis ęloquētię tribunitię potestatis facundię: q̃to tu ore patris laudes tonas:
q̃to fratris amas: q̃tus ipoetica es. O magna fęcunditas animi quēadmodum friēm quoq;
imitareris excogitasti: sed bęc quis possęt itrepidus extimare subiturus igenii tui iuditiū
pręsertim lacessitum. Neq; eim similis ē conditio publicantium & noiatim tibi dicantiū.
Tum possem dicere qd ista legis iperator bumili uulgo scripta sunt agricolaꝝ opificum
turbę deniq; studioꝝ ociosis quid te iudicē facis: quia banc operam cum dicerē nō eras in
boc albo: maiorem te sciebam quam ut descensurum buc putarem. Pręterea est quędam
publica etiam eruditoꝝ reiectio: utitur illa: et M. Tullius extra oēm ingēs italiam positus
etq; miremur per aduocatum defendit nec doctissimum ōnium Persium boc legere uolo
Lelium Congium uolo. Q si boc Lucilius qui primus condidit stili nasum legerit quasi
abusionem & uituperationē reputabit: primus enim satyricum carmen conscripsit i quo
utiq; uituperatio uniuscuiusq; continet. Nasum autē dixit quasi uituperationis signū uel
maxime naso declarandum dicendumq; ē: si aduocatum sibi putauit Cicero mutuandum
presertim cum de re publica scriberet quanto nos cautius ab aliquo iudice defendimur.

LEFT

Historia Naturalis,
Pliny,
(1469)

Published in Venice by Johann of Speyer, this is the first printed edition of Pliny's encyclopaedic guide to the natural world (see overleaf).

1 Early editions of Pliny's *Historia naturalis* by the Speyer brothers and Nicolas Jenson

The first printed botanical book that we know dates from 1469: the text is Pliny's *Historia naturalis* (see also pages 23–25). Culturally, Europe was entering the Renaissance, and classical sources were valued extremely highly for their connection with a rich past and their assumed depth of authority. *Historia naturalis* was one of the first ancient classical texts to appear in print, and is still in print today due to its significance as a source, according to his nephew Pliny the Younger, 'as full of variety as nature itself'.

Johann and Wendelin of Speyer (*fl.* 1468–77) made this first printed edition, termed by scholars the *Editio Princeps*. The brothers emigrated to Venice from Speyer in Germany, earning them their sobriquet, and were granted an exclusive right to print books in the city by the city council, dated 18 September 1468, covering the next five years. They would become the first printers to number pages with Arabic numerals, a convention we now take for granted.

Although the text is printed, the book itself was illuminated by hand. With gilding, surrounds and historiated key capitals, surrounded by twining plants in a style called Italian 'white-vine', this printed book conveys the illusion of being a manuscript, indicating the wealth of the intended purchasers.

Another early *Historia naturalis* was printed in 1472 in Venice by the printer, punch cutter and publisher Nicolas Jenson (*c.* 1420–80). Jenson was born in France and became Master of the Royal Mint, but was sent to Germany by the king to master the new technology of printing. When his master died in 1461, Jenson relocated to Venice, already a major heart of publishing, and the copy held at Reading University is one of the eleven surviving of those made by his workshop. Jenson and the Speyer brothers were among those who developed a clean Roman typography designed to look like the elegant handwriting favoured by the new humanist scholars.

These very early printed books, known as *incunabula*, closely resembled illuminated manuscripts. In fact they were explicitly printed to appear to have been handwritten, as that was the desirable existing form. The text was printed on a press, but then other skilled workers took over, as decoration was added by hand, such as beautiful page edgings or large ornamental capitals, depending on the purchaser's taste. These early editions of Pliny would have given their readers a sense of connection with the classical world and the pleasure of feeling that they were reading from an established authority.

Judging by the number of printers who undertook an edition of *Historia naturalis* alone in the late fifteenth century, the experience was highly in demand: humble Brescia, in modern-day Lombardy, Italy, had hazarded six different editions, before 1500. After 1500, even more editions were made – Paris issued thirteen; Lyons nine, of which one, dated 1562, was in native French; Cologne and London three apiece. The first Pliny in English, in contrast, was translated from the French edition. And the editions became smaller; no longer were the books all large folios (22 x 33cm/8½ x 13in), but smaller quarto (25 x 20cm/10 x 8in) and octavo sizes (15 x 23cm/6 x 9in), and even hand-sized pocket duodecimos (13 x 20cm/5 x 7¾in).

However beautifully presented, these first printed books were not illustrated. But publishers were keenly aware of the possibilities of combining image and text. Prior to this, such illustrations had been painstakingly hand-copied, limiting their accessibility and accuracy. For a short time books continued to be printed with hand illustrations added, but printers quickly realized that mechanizing illustration would be a further leap forwards. They turned to woodcuts, permitting much faster production of botanical images. Ironically, the art form would be plagued by a problem first identified by Pliny himself in Roman times: the issue of debatable plant illustrations produced by those who had not seen the plant itself, but were themselves copying from other images.

2 The first botanical woodcuts

In 1471 Albrecht Pfister (*c.* 1420–66) from Bamberg in northern Bavaria put a printed woodcut into a book for the first time – a dialogue between a ploughman and his wife – and rapidly the technique was adapted for religious images and natural history alike. Botanical images flourished: flowers appear in the printed capitals of the Augsburg Bible, printed by Günther Zainer (d. 1478) around 1475. Also in 1475, the first genuine botanical illustration appeared, also in Augsburg, in an edition of *Das Buch der Natur*

BELOW

***Das Buch der Natur*, Konrad von Megenberg, (1481)**

Megenberg's compilation of natural history, published in Augsburg, was the first book to contain printed botanical illustrations. Here the index shows the eclectic range of subjects covered.

BELOW

Ortus Sanitatis, Peter Schöffer, (1485)

The first herbal printed in a modern language, this image of medicinal herb borage, 'Borago', shows the plant's star-shaped flowers. Annotation can be seen on the right-hand page, showing the book was well-used.

BELOW

Ortus Sanitatis, Peter Schöffer, (1485)

Schöffer was a shrewd businessman as well as a skiful printer and included many Northern European plants such as these root vegetables.

Raffanus retich Capitulum.cccxxix.
Affanus latine et grece. Die meister sprechen gemeynliche
das diser retich heyß vnd drücken sy an dem dritten grad.
Retich ist glich an syner wurtzeln den ruben vñ ist vns auch
wole bekant. Retich gessen nach dem nacht essen dauwet wole die
kost vnd macht den mag⁹ warm. Aber er macht ein bösen stinckende
adem wan man balde darvff slaiffen gait.

Retich distillieret zů waßer. diß waßer ist fast gůt stranguiriosis
das ist die mit noit netzen droplingen. Auch sunderlichen dienet diß
waßer wole calculosis das ist die den steyn haben in der blasen vnd
lenden. Item retich ist den frauwen schade die kinder dragen wan
er benymt die entphangen geburt vnd brengt den frauwen mēstruū
vnd dry(e)tschoß secūdinā. das ist das fell da das kint in gelegen ist in
můter lybe. Der safft von retich ist gůt gestrichen vff alt schaden
oder wo fuil fleysch wechßet das verzert den safft vnd macht frisch die
wunden. deß glichen thut auch das pulůer von retich. Item retich
safft mit honig gesotten mit wenig eßig vermengt das durch geslagē
ist eyn gůt dranck widder das feber quartan vnd widder bestoppūg
der myltz.

RIGHT

Das Buch der Natur, Konrad von Megenberg, (1481)

This coloured woodcut appeared at the head of the chapter on trees, and a grapevine is clearly recognizable at its centre.

OPPOSITE LEFT

Ortus Sanitatis, Peter Schöffer, (1485)

Brionia is a flowering plant in Cucurbitaceae, the squash family, with curling tendrils; it was recommended in early herbals as a painkiller.

OPPOSITE RIGHT

Ortus Sanitatis, Peter Schöffer, (1485)

The Mandrake, or 'Madragora', shown with a homunculus growing at its root, to illustrate the lore that the plant would 'scream' if picked.

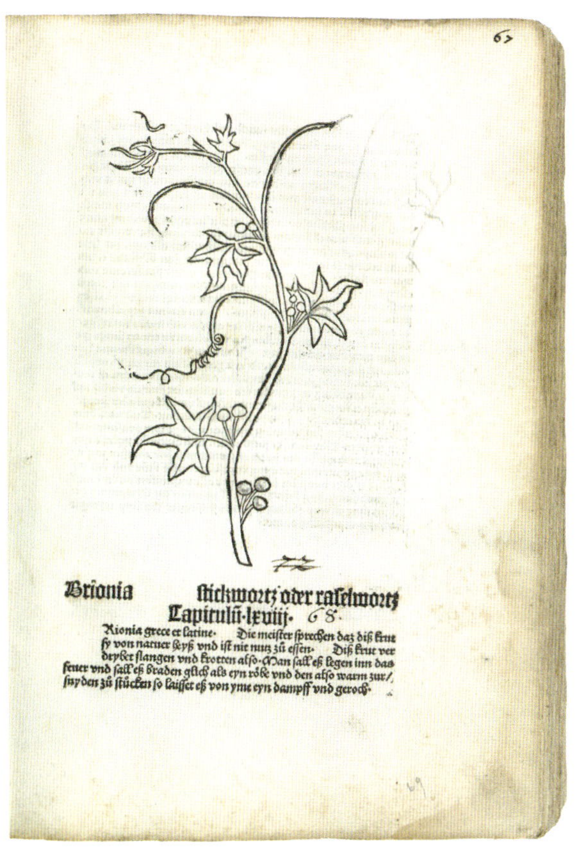

67

Brionia stickwortz oder raselwortz
Capitulũ·lxviij· 68·

Rionia grece et latine· Die meister sprechen daz diß krut sy von natuer heyß vnd ist nit nutz zů essen· Diß krut ver dryber slangen vnd krotten also· Man sall es legen inn das feuer vnd sall es braden glich als eyn röbe vnd den also warm zur/ snyden zů stücken so laisset es von yme eyn dampff vnd geroch·

Platearius disser rynden als groiß als dry heller gewicht gebak ten fur die schemde der frauwen brenget menstruũ vñ dryber vß das dot kynt· Diß rynden gestoissen zů puluer vnd genutzet mit ey nem clistier machet slaiffen vnd rüwen fur alle ander kunst· Item diß würtzel gesotten in wyn vñ vff das gegicht geleyt der gliedder ist den wetdum stillen·

·Mãdragora· Capitulum

Mandrago Die mei lichen daz selbe dogent mit vmb beschribe ich wan als du geho pitel fur dissem·

·alrun·Fraw· ·ccxlviij·

ra mulier latine· ster sprechen gemey diß alrun habe die der ersten vnd dar nit meen dar von ret haist in dem ca·

(*The Book of Nature*) by the fourteenth century author Konrad von Megenberg (1309–74) a work promising a complete natural history, printed by Johannes Bämler (1430–1503).

This innovation was extremely significant, but the promise of the new technology was not yet complete: peacocks and storks are recognizable in *Das Buch der Natur,* but the images of plants feel less successful, and only a lily of the valley and a grape vine seem clearly secure in identification. There were several issues: woodcuts wore down during printing, but also there was enormous interest in printed texts, so that printers would employ anyone who could meet their time frame, whether they knew first-hand what they were engraving or not. Perhaps from these images what we can measure is the growing thirst for knowledge about the natural world, far outstripping what printers were able to produce.

The struggle of the early woodcut makers in print workshops is in strong contrast to the beauty of plants observed in natural detail by Renaissance artists of the same moment. For example, books of hours painters recorded in beautiful, vivid detail the flowers of the wayside. *The Horae beatae Mariae ad usum Romanum* of Paris, 1524, shows extraordinarily lifelike cornflowers, bluebells and violas, alongside wild strawberries that seem almost pickable. When printing came to Paris in 1470 the city's workshops soon became specialists in printed books of hours, which appeared with the distinct densely floral borders of a painted original.

BELOW
Grimani Breviary,
unknown artist,
(*c.* 1490–1510)

This delicate illumination shows a recognizable rose bush in full summer flower, as well as white lilies, symbolic of purity.

3 The coming of the herbals: medicine for the masses

Printing helped give a strong push to the genre of herbals, guides to medicinally active plants and their uses. Before printing, herbals had been the preserve of monasteries, and were compilations from existing sources, especially from classical antiquity, often via the Arabic world, such as Galen, Hippocrates and Dioscorides (see Chapter 1). There was no swift change from this piecemeal approach, and classical authors remained a central source, individual print-shop owners often picking and choosing elements that seemed to work well for the target readers and purchasers. Perhaps because such books were often intended for those without university education, they were some of the earliest to be translated into the vernacular. However, the vast majority of early herbals were also unillustrated, involving only descriptions of useful plants to help practitioners find appropriate treatments.

The first printed herbal appeared in 1477 in Naples; *De viribus herbarum carmen*, (*On the Powers of Herbs*) published by Arnaldus de Bruxella, was a didactic poem attributed to Macer Floridus, describing the medicinal qualities of seventy-seven herbs. Much of the detail derives from the contemporary Muslim medical scholars in Sicily who translated Arabic texts into Latin. The rhyming scheme was added intentionally as a mnemonic device to help physicians, midwives and apothecaries memorize the recipes and treatments listed, but it was not until a new edition was made in Geneva in 1495 that illustrations were finally added. A plantain is distinctly recognizable from its long furry flowerhead and rosette of leaves, but other plants are not as clearly depicted. Those who wanted to identify plants from illustrations would have longer to wait.

In 1481 in Rome, for the first time, a printer made a herbal with printed illustrations of plants in it, an edition of the much-reproduced pseudo-Apuleius' *Herbarius* (see also page 25). This first printed copy was made by Johannes Philippus de Lignamine (b. 1420), the first Italian book printer; as Pope Sixtus the IV's personal physician, his close relationship to the papal court enabled him run a print shop in Rome, mainly concerning itself with religious texts.

The organization of the book suggests it was intended for practical use. In Dioscorides, the entries were set out alphabetically, but here herbs are presented in order of the diseases they are intended to treat. The late-fifteenth-century illustrations were very simplified. They were woodcuts of the much-copied drawings in the original texts, which had themselves been reproduced over and over again. Small inaccuracies had crept in and been often repeated. But this book was a landmark, nonetheless, in signposting the change that was coming.

At this time, the reader surely could not have relied on the book for identification. It has been suggested that the images acted less as identification aids and more as an index for users to quickly find their place in the (large) book. Despite the frustratingly schematic images, printed and illustrated herbals as a concept rapidly caught on and became one of the best-selling categories of early printed books: botanical knowledge was becoming a profitable business.

4 The print shop of Peter Schöffer, and the flourishing of book piracy: the market grows

German printer Peter Schöffer (*c.* 1425–*c.* 1503) began as a calligrapher in Paris before serving as foreman to Gutenberg himself in Mainz, proving a wily innovator in the field of the illustrated book. When his former master was declared bankrupt, Schöffer moved with Gutenberg's confiscated press to its new owner, the goldsmith and moneylender Johann Fust, and continued to operate it. Schöffer has been described as a 'canny businessman' and scholars surmise from textual evidence that he amassed his own collection of manuscript sources to use when compiling printed works, a rich treasure for a printer.

The first botanical book Schöffer published was the 1484 *Herbarius latinus* (*Latin Herbal*), as its title suggests, in Latin, with German names appended to each botanical entry. In it, 150 plants appeared, all of central northern European origin, although the woodcuts made are not useful in terms of actual identification. The book also contained information on mineral and animal sources of medicines, is clear on the medical uses of cannabis, and possibly shows the preoccupations of the time by having a whole chapter on laxatives. Despite its limitations, the book was extremely popular, translated rapidly

xxxj

Brionia Brionich oder
wilde weinreben.
Brionia est calide et sicce complexionis scz tota her
ba scz folia fructꝰ ⁊ radix ⁊ habent virtutem abster-
gendi et subtiliandi ⁊ dissoluendi ideo valent in duri
cie splenis faciendo emplastrū ex eo et radice altee et
ficubus cū aqua decoquendo cum axungia porci ⁊ lo
co indurato ipsius splenis applicando vel alio mē-
bro indurato ⁊ valebit. Itē cū succo brionie abradunt

into Dutch and Italian, having eleven editions before 1501. It was also, very significantly, much pirated; for example, the *Passau herbarius* (*Passau Herbal*), made by Johann Petri, in Mainz in 1484, which added distinctive local names for plants, showing an increasing demand for such information; or that made by Johann and Conrad Hist, also 1484, considered to be the rarest of all herbals, which is a complete copy of Schöffer's except that, like Petri's, all the images are back to front – a consequence of the blocks having been made from tracings of the originals.

Schöffer's next publication was the *Ortus sanitatis* or *Garten der Gesundheit* (*Garden of Health*) of 1485, which is significant in being the very first printed botanical book to appear in a modern language. Again, Schöffer showed his skill in obtaining a text – the book was written by an enthusiastic amateur. The compiler, anonymous as they remained, had realized that many of the herbs most highly recommended by the ancients did not actually grow in Germany, and claims in his preface to have set out on a study tour through the Balkans, Crete and Palestine, returning via Alexandria.

Of the 400 or so plants discussed in the text, 370 are illustrated. For the first time, some plants such as the aquilegia, bearded iris and oat are instantly recognizable from their line portraits in the text. A hop twines its way back on itself, giving an immediate sense that its observer has met it in the green. Simple shading is used to increase the sense of three-dimensionality. Some of the pictures are less successful, but nonetheless the book was such a hit that thirty-eight versions of it appeared before 1500, including translations into Flemish, French and Low German.

Judging by the texts that were the biggest sellers in the earliest years of printing, Europe had a measurable thirst for knowledge about medicinal plants. Physicians, pharmacists, midwives, herbalists and barber-surgeons might all have had an interest in owning a copy of such a book. This confluence of factors, including the advent of the printing press and increased interest in the natural world, fuelled the growth of botanical printmaking. With the printing press, botanical images could, in theory, be reproduced with greater precision and distributed much more widely. The book stopped being a resource only available to the wealthiest of patrons or to monks or students in a library. Now, knowledge was more widely available. As the market grew, so did the expectations of purchasers, and the woodblocks grew larger in relation to the text, so that by the 1560s, Antwerp printer Christoper Plantin's (1520–89) accounts show a single book's illustrations cost two-thirds of the edition's entire budget.

The precise attention and keen eye of the Northern Renaissance, and the need for greater knowledge of the natural world, began to shape a much stronger interest in what plants actually looked like. Printers began to realize they had more chance of selling a book if the illustrations enabled the reader to find their way to a medicine.

The time involved in making such accurate plates, though, was considerable, as it might stretch across seasons and years, and involve collecting from different places and regions. For this reason, there quickly developed a trade in reselling used plates to be recycled with new text. This busy industry suggests the range of books that proliferated – from lasting contributions to botanical science, right down to pirated texts and images sold much more cheaply from unscrupulous printers keen to capitalize on a growing and valuable market. Indeed, scholars have been able to quickly discern if an early book is a pirate edition by looking at the plates – publishers would copy the illustrations by tracing them, then rubbing them onto a woodblock, leaving them in tell-tale reverse in the final printing if the plates had been pirated in this way. However, publishers also bought and sold the plates themselves; carved into hardwoods, often from fruit trees such as apple and pear, they could still be used to print even as they showed signs of long use.

It is striking that at this time painters were also focusing in more detail on the botanical structures of plants and flowers. The edges of fine and costly illustrated manuscripts, such as the *Grimani Breviary* (*c.* 1515) and the *Book of Hours* created for Anne of Brittany (1477–1514), offer insight into both the artists' search for the materiality of plants and their deep understanding of the natural world. Plants and flowers are also depicted in such manuscripts for their symbolic significance, representing virtues or religious themes, such as purity or humility. In this way they serve as both decorative elements and conduits for deeper layers of meaning; but also, as a way into the picture – as anyone who has begun looking at a religious painting by starting with the flowers in the foreground will know. The realistic depiction of flowers in religious art aims to assure the viewer of the reality of religious experience, of the

OPPOSITE LEFT

Briony, from the *Passau Herbal*, (1484)

A pirated version of Schöffer's Latin herbal can be compared to the illustration from Schöffer's original on page 59.

OPPOSITE RIGHT

***Ortus sanitatis*, (1485)**

The frontispiece of *Ortus sanitatis* (*Garden of Health*), printed by Peter Schöffer, an expansion of previous texts and the first in a modern European language.

physical nature of Christ's incarnation in flesh and of God's presence in the everyday world. Whether for spiritual or economic reasons, botanical eyes were paying greater and greater attention to the natural world.

5 Otto Brunfels, Hans Weiditz and the rise of observational illustrations

Otto Brunfels (1488–1534), from printing's founding city of Mainz, made a significant contribution to botany, particularly through his *Herbarium vivae eicones* (*Pictures of Plants from Life*). Brunfels began his working life as a minister, came into conflict with theologians Luther, Erasmus and Zwingli, and ended his days in Bern in the Swiss confederacy, where he was the city physician.

His contribution was simple: he recorded his own observations and critiqued the received wisdom of Dioscorides and Pliny, trying to correct or confirm where he could. He aimed to produce an account of the world that valued the knowledge of the present over the past, and his books show images full of life and growth suggesting the vivid reality of living plants.

Brunfels championed the use of technical Latin terms for parts of the plant, giving an official language to plant identification and classification. His biggest preoccupation was verification. He went to some trouble to collect vernacular names for plants with medicinal properties and then carefully and studiously identified the Latin and Greek sources where they were discussed.

His work was made all the more important via his collaboration with Hans Weiditz (1495–1537), who made the woodcuts for his books. This matchmaking was probably done by their publisher, Johann Schott of Strasburg. Weiditz came from an artistic family, his father described as a sculptor in the guild records of Freiburg, alongside Hans Baldung who had studied with the celebrated German printmaker and painter, Albrecht Dürer. Weiditz eventually would be part of this flourishing art scene, which also included Hans Holbein before he moved to England to become the court painter to Henry VIII. Weiditz later lived in Augsburg in Bavaria, which dominated banking in the German region, as well as being the most important postal centre in the Holy Roman Empire.

Weiditz's images are deeply sympathetic to the growing habits of real plants; his pasqueflower capturing featheriness of that plant, with its bright flowers shown in full bloom as well as bud. His narcissi bulbs are thick with roots sent down to stabilize the plant and take moisture from the soil. They are a joy to behold. But they are also accurate in a way that assures the reader that they have been drawn from life.

RIGHT
Otto Brunfels,
Hans Baldung,
(*c.* 1534)

Brunfels is portrayed here by Baldung, a friend and colleague of Albrecht Dürer, who specialized in woodcuts for books and characterful portraits.

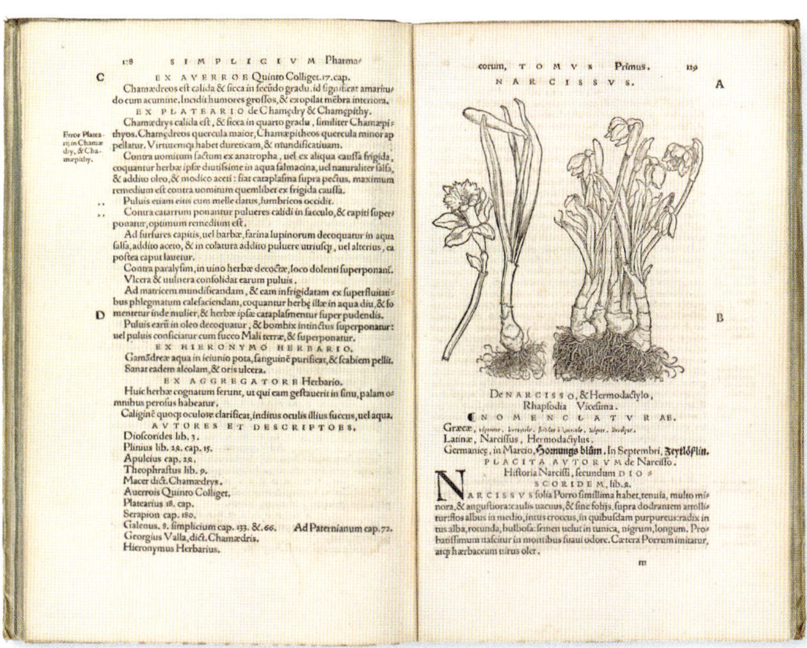
128 SIMPLICIVM Pharma-

C EX AVERROE Quinto Colliget. 17. cap.

Chamædreos eſt calida & ſicca in ſecũdo gradu. id ſignificat amaritudo cum acumine. Incidit humores groſſos, & exopilat mẽbra interiora.

EX PLATEARIO de Chamędry & Chamępithy.

Error Platearij in Chamædry, & Chamæpithy.

Chamædrys calida eſt, & ſicca in quarto gradu, ſimiliter Chamæpithyos. Chamędreos quercula maior, Chamæpitheos quercula minor appellatur. Virtutemq; habet dureticam, & mundificatiuam.

Contra uomitum factum ex anatropha, uel ex aliqua cauſſa frigida, coquantur herbæ ipſæ diutiſsime in aqua ſalmacina, uel naturaliter ſalſa, & addito oleo, & modico aceti: fiat cataplaſma ſupra pectus. maximum remedium eſt contra uomitum quemlibet ex frigida cauſſa.

Puluis etiam eius cum melle datus, lumbricos occidit.

Contra catarrum ponantur pulueres calidi in ſacculo, & capiti ſuperponatur, optimum remedium eſt.

Ad furfures capitis, uel barbæ, farina lupinorum decoquatur in aqua ſalſa, addito aceto, & in colatura addito puluere utriuſq;, uel alterius, ea poſtea caput lauetur.

Contra paralyſim, in uino herbæ decoctæ, loco dolenti ſuperponant.

Vlcera & uulnera conſolidat earum puluis.

Ad matricem mundificandam, & eam infrigidatam ex ſuperfluitatibus phlegmatum calefaciendam, coquantur herbę illæ in aqua diu, & fo-

D mentetur inde mulier, & herbæ ipſæ cataplaſmentur ſuper pudendis.

Puluis earũ in oleo decoquatur, & bombix intinctus ſuperponatur: uel puluis conficiatur cum ſucco Mali terræ, & ſuperponatur.

EX HIERONYMO HERBARIO.

Gamãdreæ aqua in ieiunio pota, ſanguinẽ purificat, & ſcabiem pellit. Sanat eadem alcolam, & oris ulcera.

EX AGGREGATORE Herbario.

Huic herbæ cognatum ferunt, ut qui eam geſtauerit in ſinu, palam omnibus peroſus habeatur.

Caliginẽ quoq; oculorũ clarificat, inditus oculis illius ſuccus, uel aqua.

AVTORES ET DESCRIPTORES.

Dioſcorides lib. 3.
Plinius lib. 28. cap. 15.
Apuleius cap. 28.
Theophraſtus lib. 9.
Macer dict. Chamædrys.
Auerrois Quinto Colliget.
Platearius 18. cap.
Serapion cap. 180.
Galenus. 9. ſimplicium cap. 133. & 66. Ad Paternianum cap. 72.
Georgius Valla, dict. Chamædris.
Hieronymus Herbarius.

corum, TOMVS Primus. 129

NARCISSVS. A

B

De NARCISSO, & Hermodactylo, Rhapſodia Viceſima.

NOMENCLATVRAE.

Græcæ, ...

Latinæ, Narciſſus, Hermodactylus.

Germanicę, in Marcio, Hornungs blům. In Septembri, Zeytlöſlin.

PLACITA AVTORVM de Narciſſo.

Hiſtoria Narciſſi, ſecundum DIOSCORIDEM, lib. 4.

NARCISSVS folia Porro ſimillima habet, tenuia, multo minora, & anguſtiora: caulis uacuus, & ſine folijs, ſupra dodrantem attollitur: flos albus in medio, intus croceus, in quibuſdam purpureus: radix intus alba, rotunda, bulboſa: ſemen uelut in tunica, nigrum, longum. Probatiſſimum naſcitur in montibus ſuaui odore. Cætera Porrum imitatur, atq; hærbaceum uirus olet.

m

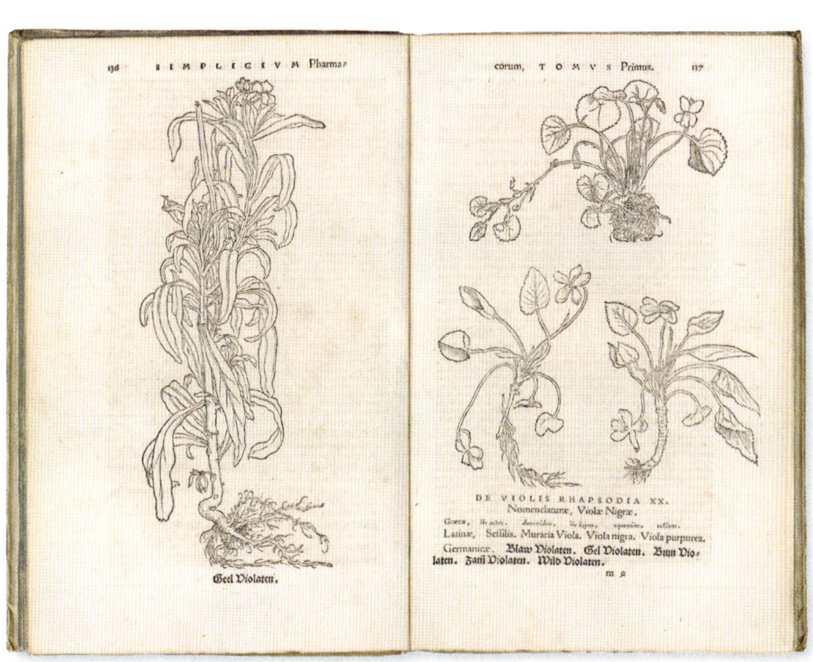
136 SIMPLICIVM Pharma-

Geel Violaten.

corum, TOMVS Primus. 137

DE VIOLIS RHAPSODIA XX.

Nomenclaturæ, Violæ Nigræ.

Græcæ, ...

Latinæ, Seſſilis. Muraria Viola. Viola nigra. Viola purpurea.

Germanicæ, Blaw Violaten. Gel Violaten. Brun Violaten. Zam Violaten. Wild Violaten.

m 4

LEFT

***Herbarium viva eicones*, Otto Brunfels, (1530)**

Narcissi (top) and violet flowers (bottom), showing detail of blooms and natural growth in a clump, in woodcuts by Hans Weiditz the Younger; with text showing the diversity of the plant names in various languages.

6 Leonhart Fuchs and the first botanical bestseller

Leonhart Fuchs (1501–66), a Basel physician and botanist, published *De historia stirpium commentarii insignes* (*Remarkable Commentaries on the History of Plant Races*) in 1542. Fuchs studied medicine, then taught university medical students in central Germany, creating a botanic garden at Tübingen for practical teaching. He was a fierce character: his first book, in 1530, was called *Errata recentiorum medicorum* (*The Errors of Modern Doctors*) and he resisted the trend towards Arabist medicine based on ideas originating from Sicily, preferring a strict return to classical Greek learning. He had little time for pharmacists, herbalists or apothecaries, whom he described as 'uneducated men and stupid little women'. He corresponded with botanical experts elsewhere, such as the Pisan Luca Ghini (see also page 74), but he fell out categorically with Pietro Andrea Mattioli, whom he would call a 'pride-swollen Italian' (see also page 76).

De historia stirpium was notable for its detailed and lifelike woodcut illustrations of almost 500 different plants. The printmaking was so significant that Albrecht Mayer, the artist who made the drawings, Heinrich Füllmaurer, the draughtsman who transferred the drawings to the blocks, and Vitus Rudolph Speckle, the woodcut engraver, all appear together in a charming frontispiece illustration. Mayer is seen in the illustration observing a plant from life, indicating the high standards that the work set itself. This was the first time that such a portrayal had appeared of the named artisans who created the pictures.

Fuchs had mostly re-used Dioscorides' text, but the illustrations he commissioned confirmed the high value of accurate representative images. His book had more than twice as many images as Brunfels', was printed on much better-quality paper, had larger pages and was laid out with a far more generous sense of space. A glorious spiny acanthus fills the page. This text was also the first to show several species new to Western eyes, such as the maize plant and the chilli pepper, brought from the Americas by Europeans. The title page shows the print mark of the work's Basel-based publisher, Michael Isingrin: a palm tree, with the Latin inscription 'PALMA ISING'. Fuchs's book, it would be fair to say, was a bestseller, appearing thirty-nine times during his own lifetime alone, in Latin but also Spanish, French, German and Dutch, and the woodcuts were of such a quality as to continue appearing in books for another 300 years – the print blocks repeatedly reused, sold on by Isingrin to fellow printers in Paris, London and Antwerp.

However Fuchs' work was not finished after *De historia stirpium*, and he set himself the task of compiling a much larger encyclopaedia. This was never finished, but its manuscript is now known as the *Codex Fuchs* (*Codex Vindobonensis Palatinus*). This illustrated for the first time many more plants newly arrived from the Americas, such as marigolds, pumpkins and

OPPOSITE

The Large Piece of Turf,
Albrecht Dürer,
(1503)

This watercolour is so detailed the species pictured can easily be identified: dandelion, coltsfoot, plantain and germander speedwell, growing in a natural wild fashion.

BELOW

Lily,
Albrecht Dürer,
(1526)

This watercolour study shows the stamens of the lily, heavy with rust-red pollen.

tobacco, which Fuchs named *Nicotiana*. He called his critics 'long-nosed nitpickers', but most of them beat him into print with the novelties he could have shown first, and the work has still never been published. In tribute, nonetheless, the *Fuchsia* was eventually named after him.

Today, scholars increasingly understand that these sorts of herbals were not simply treasured, but were working sources, and were often annotated, added to and even equipped with pressed plant specimens between their pages. In the past, such heavily-used volumes would have been assessed by librarians and antiquarian book dealers as 'damaged' and considered in poor condition; but in recent years there has been a pivot in appreciating how such evidence reveals the way the books were used, and the true state of knowledge in the field. For example, in a copy of Fuchs's *De historia stirpium* owned by Glasgow University, which belonged to Lord Fauconberg, English common names have been added, leaves have been pressed between the appropriate pages, and cross-references written in for the *Herball* of the renowned English herbalist John Gerard (1545–1612). In another such example, in a *c.* 1560 Venetian drawing album held in the British Museum, and once owned by Anglo-Irish naturalist and collector Hans Sloane, we can see how the artist carefully copied the outlines of a Solomon's seal exactly as it is shown by Fuchs; to finish the image, the artist added a few genuinely pressed leaves from the same plant, still shapely after 500 years. The combination of study and real specimen gives us an insight into the knowledge for which this reader searched, matching the world of the book with the plant itself.

OPPOSITE

***De historia stirpium*, Leonhart Fuchs, (1542)**

Papaver rhoeas, the poppy, called here 'Papaver erraticum', showing red flowers on long stems in a hand-coloured engraving, opposite Fuchs' Latin text.

BELOW

Leonhart Fuchs painted by Heinrich Füllmaurer, (1541)

Fuchs appears in this oil painting wearing the fur hood and sober dress of a Lutheran doctor, holding a plant thought to be common fumitory.

About 650 botanical works were published in Europe in the sixteenth century, giving us a sense of a rich and enthusiastic culture of use. Increasing urbanization and diseases of overcrowding such as cholera and plague created more customers for medical herbalists, although we might also imagine that increasing disposable wealth meant the new customers also arrived with higher expectations of being helped. William Turner (1508–68) in England was typical of the latter half of the century in pursuing some sort of standardization; he listed the names of the plants he included in Greek, English, French and Dutch, including a range of common names, hoping to help apothecaries actually identify the plants in question.

Such books would have given their owners a sense of participating in a new culture, a feeling of connecting to precious knowledge gathered by ancient authorities, but repackaged in the newest style. The book's readers would have felt part of a community aiming to widen existing networks, practice and expertise, using such books for teaching and study, perhaps linking to scrutiny of actual specimens in the real world, for example in another newly developing area: the botanic garden.

514 PLANTARVM HISTORIAE CAP. CCXIII.

C potu, cœliacis, fœminarũ fluxionibus, morbo regio, & phalangiorũ morsibus prodest. Oedemata illitus reprimit. Cortex eadem quæ fructus potest. Decoctum foliorum cum uino potum, lienem minuit. Ad dolorem dentium in collutionibus salutare est, & mulierum fluxionibus in desessionibus. Aptè his in quibus pediculi & lendes abundant circunfunditur. Cinis lignorum eius appositus profluuia ex utero supprimit. E trunco eius nonnulli potorios calices fabricant, quibus in lienosis pro poculo utuntur, quò datus in ijs potus proficiat.

Calices potorij ex trunco tamaricis.

EX GALENO.

Myrice prodest admodũ lieni indurato, decoctis cum aceto aut uino radicibus, siue folijs, siue extremis ramulis. Sanat uerò etiam dentium dolores.

EX PLINIO.

Lenæus sanari ea carcinomata in uino decocta tritaque cum melle dicit. Ad lienem præcipua est, si succus eius expressus in uino bibatur. Adeoque mirabilem eius [illegible] pathiam cõtra solum hoc uiscerum faciunt, ut affirment, si ex ea alueis factis bibant sues, sine liene inueniri. Et ideo homini quoque splenetico cibum potumque dant in uasis ex ea factis. Lignum & flos & folia & cortex in eosdem usus adhiberi, quanquam remissiora. Datur sanguinem reijcientibus cortex tritus, & contra profluuia fœminarum, cœliacis quoque. Idem tusus impositusque collectiones omnes inhibet. Folijs exprimitur succus. Ad hæc eadem & in uino decoquuntur: ipsa uerò ad [illegible] melle gangrenis illinuntur. Decoctum eorum in uino potum, uel impositum cum rosaceo & cera sedat. Sic & epinyctidas sanant. Ad dentium dolorem [illegible] decoctum eorum salutare est. Radix ad eadem, similiter & folia. Hæc [illegible] quæ serpunt imponuntur cum polenta. Semen drachmæ pondere [illegible] langia & araneos bibitur. Cum altilium uerò pingui furunculis imponitur. [illegible]

D & contra serpentium ictus, præterquam aspidum. Necnon morbo regio, [illegible] si, lendibusque decoctum infusum prodest, abundantiamque mulierum sistit. Cinis arboris ad omnia eadem prodest.

DE MECONE RHOEADE. CAP. CXCIIII.

NOMINA.

Papauer rubeũ.

ΜΗΚΩΝ ῥοιὰς Græcis, Papauer rhœas, hoc est, fluidum aut erraticum Latinis, Papauer rubeum officinis & herbarijs, Germanis à strepitu quem ludentes pueri harum folijs concauo pugno impositis, altera palma incussa ædunt, Klapperrosen/ quasi dicas crepitaculares rosas, & Glitschen/ & wild Maen/ ac Kornrosen appellant. Rhœas autem à flore, quod protinus decidit, dictum est.

Rhœas unde dictum.

FORMA.

† aliâs iunceum

Folia Erucæ, aut Cichorio similia, & incisa habet, longiora tamen & aspera. Caulem † lanuginosum, rectum, asperum, cubitalem. Florem puniceum, & aliquando candidum, similem syluestris Anemones flori. Caput oblongum, minus tamen quàm Anemones. Semen rufum. Radicem oblongam, subalbam, minoris digiti crassitudine, amaram. Ex qua deliniatione omnibus perspicuũ fit, herbam quam Papauer rubeum hodie uulgò nominant, esse Rhœada, quod sit folio Erucæ, laciniato, scabro, & longiore: caule lanuginoso, recto, aspero, cubitali: flore syluestris Anemones puniceo, nonnunquã albo, oblongo capite: semine ruso, radice longa, subalbida, minoris digiti crassitudine, gustu amara. Errant itaque qui Anemonen esse arbitrantur, quod folia eius, quæ Coriandri sunt, euidentissimè reclamant. Nos Rhœadis utriusque picturas damus. Vnius, atqueadeo primi, quod Erucæ folia habet, alterius quod Cichorij.

Papauer rubeũ nõ est Anemone.

LOCVS.

515

PAPAVER ERRATICVM PRIMVM.

Klapperrosen.

Poppy.

BELOW

***De historia stirpium*, Leonhart Fuchs, (1542)**

Cannabis sativa, in sprightly fresh green growth, in an elegant woodcut demonstrating the clear style Fuchs demanded of his artists.

392 PLANTARVM HISTORIAE CAP. CXLVI.

c disijcit & dissoluit. Præsens remedium est aphthis & ulceribus in ore serpentibus. Decoctum eius tumoribus illitum eos dissipat. Vsus etiam eius ad intertrigines, cera pudendorum, & intestinorum exulcerationes prodest.

DE CANNABE. CAP. CXLVII.

NOMINA.

Κannabis Græcis, Cannabis Latinis, Barbaris & uulgo Canapus dicitur, Germanis autem Hanff.

GENERA.

Cannabis duo sunt genera. Vna enim satiua est, quam Græci σχοινόστροφον, quod magni in uita usus sit ad robustissimos funes texendos. Germanis zamer Hanff dicitur. Altera syluestris, quam Latini Terminalem uocant, Germani wilden Hanff.

FORMA.

Satiua Cannabis. Satiua Cannabis folia fert fraxino similia, grauis odoris, caules longos, semen rotundum. Syluestris. Syluestris uerò uirgas fundit Altheæ similes, nigriores, res & minores, cubitali altitudine. Folia satiuæ similia, asperiora & nigriora. subrubeos, Lychnidi similes. Semen & radicem Altheæ similia. Eius effigiem dere nobis nondum licuit.

LOCVS.

Satiua, in locis cultis sata prouenit. Syluestris in syluis & asperis locis, teste, iuxta semitas & sepes nascitur.

TEMPVS.

Herba ad usus medicos carpitur dum maxime uiret. Semen autem eius, D autore, cum maturum est, id quod prope autumni æquinoctium accidit.

TEMPERAMENTVM.

Admodum calefacit & exiccat.

VIRES. EX DIOSCORIDE.

Satiuæ maiori copia sumptum genituram extinguit. Ex uiridi autem pressus, contra aurium dolores utiliter instillatur. Syluestris radix inflammationes mitigat. Tumores discutit, & callos disijcit. Huius cortex funibus utilis est.

EX GALENO.

Cannabis semen flatus extinguit, adeoque desiccat, ut si plusculum edatur ram exiccet. Sunt qui ex uiridi succum exprimentes, ad aurium dolores ctione, ut mihi uidetur, natos utuntur. Semen etiam cannabinum quitur, stomacho & capiti aduersatur, prauosque humores creat. Admodum facit, ideoque caput æstuoso ac medicamentoso halitu sursum misso tentat.

EX PLINIO.

Semen Cannabis extinguere genituram uirorum dicitur. Succus ex eo culos aurium, & quodcunque animal intrauerit, eijcit, sed cum dolore capitis. taque uis ei est, ut aquæ infusa, coagulare dicatur, & ideo iumentorum aluo pota in aqua. Radix cōtractos articulos emollit in aqua cocta. Item podagras miles impetus. Ambustis cruda illinitur, sed sæpius mutatur priusquam

EX SYMEONE SETHI.

Semē Cannabis comestum idem nocumentum quod Coriandrū affert: dicè enim si estur, ut illud, deliriū facit. Folia uero arida pota, ueluti farina, gis pro potione hæc farina exiccata, ebrietatem quandam hospitalem facit, ab hauriente non sentiatur. Apud Arabas enim pinsitur, subigiturue pro inebriat. Desiccat uerò semen genitale ut Caphura.

393

CANNABIS SATIVA

Zamer Hanff.

Hemp.

BELOW

***De historia stirpium*, Leonhart Fuchs, (1542)**

This edition of Fuchs' work, in the Wellcome Collection in London, has a handwritten index of English plant names added by an owner to enhance its use.

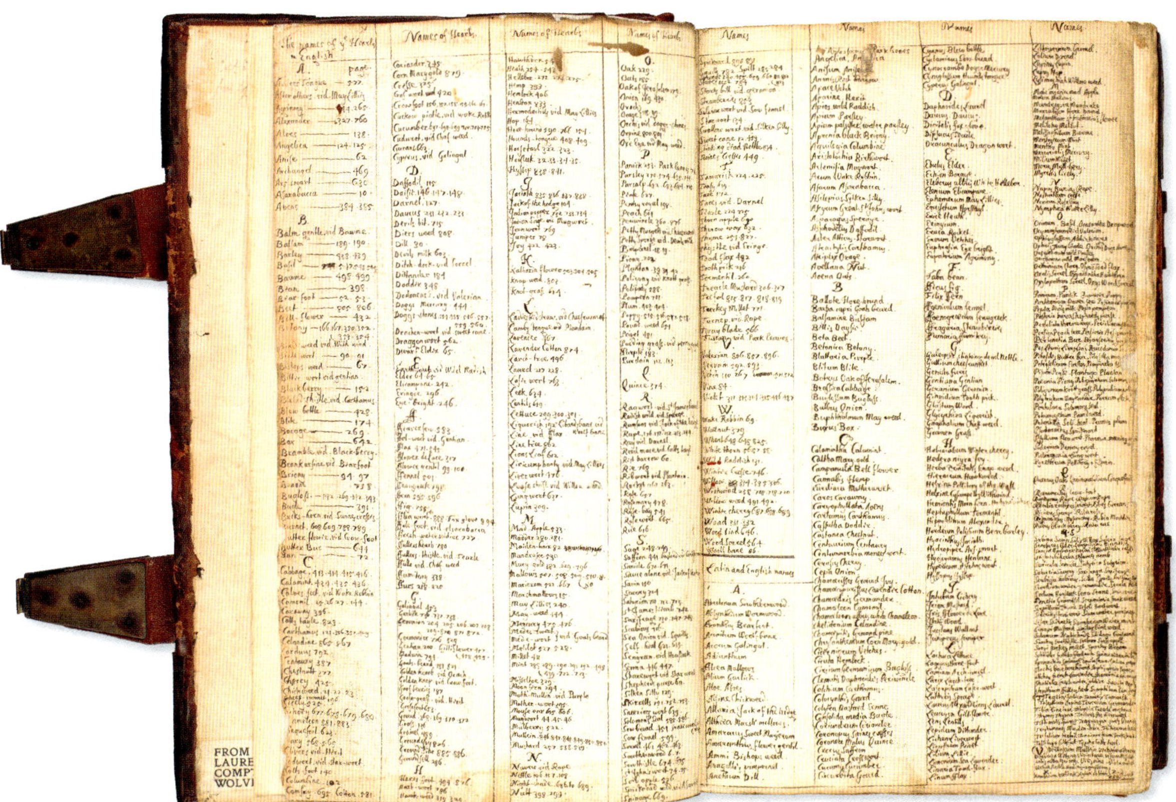

BELOW

De historia stirpium, **Leonhart Fuchs,** (1542)

Hedera helix, the common ivy, showing its curling stems and adventitious roots, seeking points of attachment as the plant climbs.

BELOW

De historia stirpium, **Leonhart Fuchs,** (1542)

The chilli pepper, 'Calechutsicher Pfeffer', with its distinctive red fruit and seeds, making its first appearance in European botanical literature.

BELOW

De historia stirpium,
Leonhart Fuchs,
(1542)

Aquilegia, or Columbine, instantly recognizable for its soft green leaves and bell-like purple flowers hanging on bending stems.

BELOW

Neue Kreüterbuch,
Leonhart Fuchs,
(1542)

Fuchs' Latin text was quickly published in German, with a much more useful index; here, mistletoe shows its characteristic globe of leaves and small white berries.

7 Luca Ghini and the botanic garden

Books were important to these early botanists, but it was increasingly recognized that knowledge from the real world of plants was essential too. Monastic gardens were recorded for medicinal purposes from the ninth century, for example, the one at the Abbey of St Gall in Switzerland established by the decree of the Holy Roman Emperor, Charlemagne, known as a 'physic garden', for which we still have plans and a plant list. By the eleventh century, most monasteries would have had an area for cultivation, such as the one at Westminster Abbey, London, said to be the oldest continuously cultivated garden in England.

However, the first explicitly botanic gardens in Europe were established in Italy, such as the one in Padua, created in 1545 by order of its then rulers, the Senate of the Venetian Republic. Venetian soils were perhaps too saline for really successful gardening, so Padua was an ideal close-by location. The garden's creator was Francesco Bonafede (*fl.* 1530s), who had lobbied for the creation of a teaching chair for medicinal plants at the University of Padua in the 1530s. Bonafede made his lectures separate from teaching sessions that were held in the garden, showing that both parts of the educational process were increasingly seen as important.

Padua was the first botanical garden to have its own guidebook, published in 1591 by a copper engraver called Girolamo Porro (*c.* 1520–1604), including a beautiful folding plan of the garden's Renaissance layout. This consisted of circles within squares within circles again, reminiscent of the patterns in the Renaissance books of emblems that had become fashionable among educated people. Similar botanic gardens were made at Pisa (1544) and Bologna (1547); others followed. The layout of these early botanic gardens, using pure geometric shapes, helped to suggest that, in entering the garden, a student could see the totality of the plant world. The garden promised the possibility of knowing everything, a complete and unified display of nature.

BELOW

Portrait of Luca Ghini, Pisa, (1490–1556)

This portrait of Ghini held in Pisa's Orto Botanico, which he created from 1544 onwards, shows him proudly holding one of his own books.

At Pisa, the making of the garden from 1544 was superintended by Luca Ghini (1490–1556), an energetic expert. Ghini had set up the Bologna course on 'simples', the early term for medicinal botanical teaching, within the medical faculty, eventually earning himself the title *professor simplicium*. He was then headhunted from Bologna by Cosimo de Medici himself; Cosimo's offer to fund a teaching garden, while Bologna hesitated, tempted Ghini's defection. Ghini worked hard to amass plants for the new Pisa garden, using contacts to secure plants from Egypt, Sicily, Syria and Crete; by 1648 a list now in the Florentine city archives shows the garden was already growing 620 different named plants.

Ghini taught his pupils outside in the garden itself, and a generation of his students included Andrea Cesalpino (*c.* 1524–1603) and Ulisse Aldrovandi (1522–1605), both significant figures discussed below. In Aldrovandi's papers, scholars would later find notes on 103 of Ghini's lectures attended by the younger man in his youth, including notes on discussions of Dioscorides. Another of Ghini's students, Gherardo Cibo (1512–1600), made paintings now in the British Library, which show herbalists collecting in the field, an almost unique glimpse into their working practices; one figure holding a plant up to the sunlight, while his colleague consults a book.

In letters to Aldrovandi, we also find reference to *Ghini's hortus siccus* or 'dried garden', comprising dried and pressed specimens of living plants. This significant innovation quickly caught on, and dry specimens began to be exchanged between experts in different cities, forming the first scientific herbaria. A herbarium emphasized that the physical reality of the plant should be the determining aspect in its identification. They became a central tool in the work library of a botanist.

8 Cesalpino and the first herbaria arranged by similarities

Ghini's herbarium is now lost, but that of his pupil and successor Andrea Cesalpino still exists, now held in the Florence Natural History Museum. Cesalpino made the next significant step forwards, for the first time organizing plants by their resemblance to one another, rather than by what illnesses they cured or simply alphabetically. Cesalpino looked for organizational unity and repeating patterns in the plant world, according to philosophical principles, and for this he is recognized as a true innovator.

Cesalpino's 1563 herbarium index appears in Greek, Latin and Italian, and is handwritten. There are 260 herbarium sheets, folio-sized, so that each holds more than one dried specimen: in its entirety it contains 768 plants, arranged according to similarities in fruits and seeds. Like Theophrastus so many years before, Cesalpino believed that a plant's ultimate goal was to produce seeds, and so the seed was the deepest identity and best 'key' to use in a classification. He also listed his reasons for discarding the methods used by others: roots were too similar and foliage varied throughout the year.

Cesalpino's volume begins with trees and moves through taller shrubs to the flowering plants. Botanical minds were striving for order: this was entirely the first time a botanist had made an arrangement like this. 'All science', wrote Cesalpino in his 1583 *De plantis libri XVI*, 'consists in the gathering together of things that are alike, and in the distinguishing and separating of the unlike.' This printed work was unillustrated; Cesalpino described the book as 'most pure, unadulterated by any images'. This would not be the first or last time that a botanist expressed doubt about illustrations.

Following Cesalpino's Aristotelian philosophical beliefs – looking for essential resemblances that indicated family identities – he extended his system further, so that this second text encompassed 1,500 plants. He divided the plant kingdom into thirty two groups, some still extant today: the grasses and the mints, for example, were each correctly seen as comprising a related 'family'. Cesalpino saw that the round, many-petalled dandelion and daisy bore resemblances to each other, and included them in a group called the *Compositae*, still recognized in modern botany.

Like Ghini, Cesalpino used contacts overseas to gather specimens of new and interesting plants – but also faced intense pressure as each new discovery further expanded what they considered to be 'the world of plants'. Cesalpino wrote of the overwhelming arrivals, 'as the proverb goes, day after day Africa yields something new'. But he also paid attention to the reality of familiar everyday plants – describing and studying the growing roots of ivy as it winds up a building, the climbing tendrils of clematis and the elegant way a fern's leaves uncurl.

9 Mattioli, the power of networks and a bestseller

It was, however, the Sienese physician Pietro Andrea Mattioli (*c.* 1501–77), who became the surprise botanical bestseller of the late sixteenth century, with *Commentarii in sex libros Pedacii Dioscoridis Anazarbei de Materia medica* (*A Commentary on Six Books of Dioscorides de Materia Medica*) published in 1544 in Venice. Ostensibly this was a translation of Dioscorides with a commentary, and Mattioli was loyal to the original, translating the ancient Greek into modern Italian, but as each novel edition appeared, Mattioli added more newly introduced plants. Many novelties were arriving at the time from Istanbul, and Mattioli became their first place of publication: for example, lilacs, which arrived with the Holy Roman Empire's ambassador to Constantinople, Ogier Ghiselin de Busbecq.

As the work expanded, Mattioli's words began to outnumber Dioscorides': the book would eventually go into sixty-one different editions, selling an astonishing 32,000 copies. Particularly noted for their illustrations are the 1562 and 1565 editions made in Venice; large pear-wood planks were carved for the woodcuts, over a hundred of which still exist, now museum pieces in their own right. They include one of the first depictions of a tulip in western art – listed (incorrectly) as a daffodil. The lychnis woodblock has its distinctively hairy leaves and tightly branching flowers, and the detail of the cutting examined up close is exquisite.

The book was mainly aimed at summarizing the medically useful plants that had been the focus of the Greek author, so that the staff of the Greek god of healing, Asclepius, with an entwined snake, appeared on the title page. But this was also the first time, especially in later editions, that such a book included some plants that lacked any known medical usefulness. This marks a move from the study of plants being purely medicinal in focus. Mattioli was a tireless but tiring man, brooking no criticism, and worked non-stop to ensure that his work, whether the best or not, became the dominant trope through which European botany would be seen. He created one of the most successful botanical networks, consistently bringing out new editions of his books and suggesting to a large group of correspondents that their libraries were incomplete without his work. But Mattioli may have had a streak of softness in him: when his teacher and colleague Ghini finally died, he wrote that the loss had carried away half of his heart.

10 Hieronymus Bock, Ulisse Aldrovandi, and attention to the local

Those interested in plants increasingly wanted to be able to identify what was growing in their own locality. A good example is the work of Hieronymus Bock (1498–1554), a minister and botanist from the Rhineland whose 1546 *Kreutterbuch,* or herbal, was without illustrations, but came with careful descriptions. Later editions, however, were illustrated, with some woodcuts taken from Brunfels and Fuchs, but also comprising 500 originals by David Kandel (1520–92).

Bock began his career as a school teacher but his interest lay in the outdoors, and he took long walks, paying attention to local wild plants. His effort focused on trying to identify these German natives with those appearing in Dioscorides. Yet Kandel's illustrations had some of the same issues we have already seen – the Oxford specialist on botanical art, Stephen Harris, has pointed out some of the woodcuts are what he terms 'imaginative chimaera'. However the fig leaf is instantly recognizable, as is the upset stomach that overeaters of its fruit may experience, fully expressed in this image.

In Italy, Ulisse Aldrovandi (1522–1605) also aimed to identify and name the local flora. He organized an expedition into the mountains in Umbria, and exchanged his knowledge with similarly passionate colleagues across Europe, most notably the Flemish physician and botanist Charles de l'Écluse (1526–1609; better known as Carolus Clusius) in Montpellier, and Jean de Brancion in Antwerp. Their most important innovation was to begin trying to standardize a system of Latin names for plants. Aldrovandi went on to start the botanic garden at the University of Bologna, the Orto Botanico.

And in London, by 1596 John Gerard had a garden on Fetter Lane in Holborn, London, which had a catalogue of over a thousand plant names, considered to be the first plant catalogue of a single garden. Gerard was the superintendent for the gardens belonging to Lord Burghley, Queen Elizabeth I's adviser and High Treasurer; he then was involved with the Barber-Surgeons' Company in the City of London, and eventually was made an official supplier of herbs and other produce to Queen Anne, in return for a grant of land next to Somerset House.

BELOW

Ulisse Aldrovandi, (1522–1605)

Aldrovandi's herbarium of almost 5000 dried specimens is still held at the University of Bologna, where he also created the medical department's first botanic garden: this is the surviving title page of the herbarium catalogue.

THIS PAGE

Herbarium catalogue of Ulisse Aldrovandi, (1550 onwards)

Eventually running to sixteen volumes, Aldrovandi's herbarium comprised dried plant specimens preserved on sheets of paper, one of the earliest survivals of such a collection.

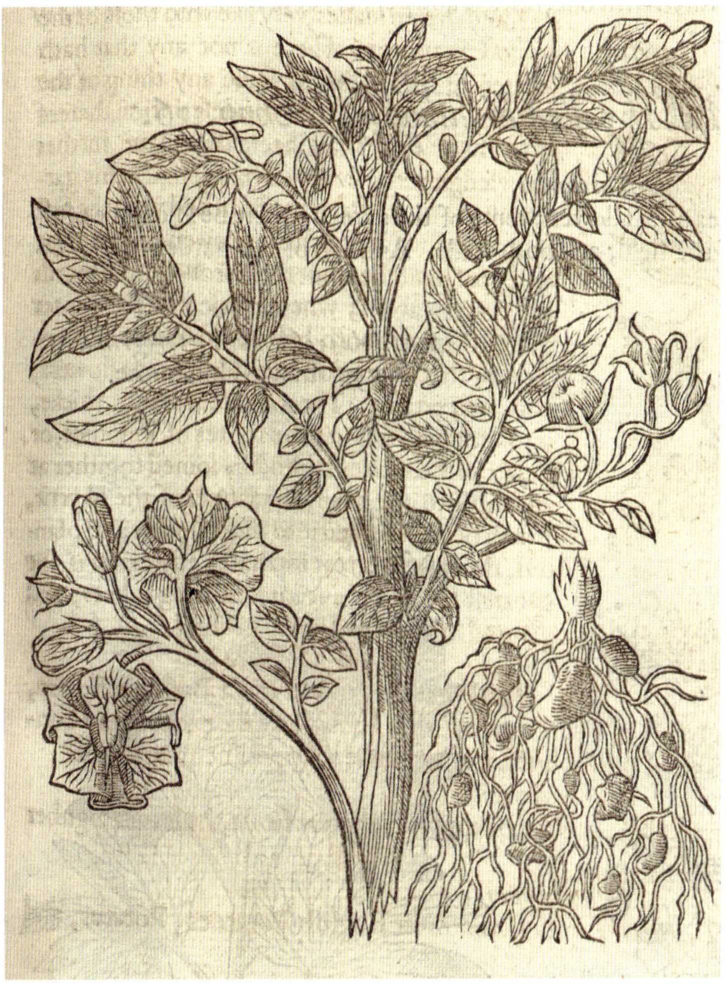

ABOVE

The Herball, or *Generall History of Plantes*, John Gerard, (1597)

Gerard is shown in the glorious Renaissance frontispiece to his herbal, alongside the first illustration of a 'Virginia Potato', recently arrived from British America.

RIGHT

***Stirpium adversaria nova*, Matthias de l'Obel, (1570)**

Published in London by Thomas Proudfoot, and organised by botanical characteristics rather than medical uses, this book was valued for its many detailed illustrations.

At the same time, John Gerard compiled his *Herball, or Generall Historie of Plantes*, printed in London in 1597. Although modern scholars have identified many significant borrowings in the text, this was a standard practice, and Gerard did include images of plants newly introduced from the Americas – such as the 'Potato of Virginia', now known as *Solanum tuberosum*, the potato.

11 Lobelius and the continuing study of plants from life

Matthias de l'Obel (1538–1616) was a Flemish physician, who was born in Lille and lived in London, Delft and the thriving city of Antwerp, which would soon gain from Venice the preeminent position as a global powerhouse. Most often known as 'Lobelius' – the Latin version of his name that would have appeared on the front page of books – his most significant publication came in 1570 with the herbal, *Stirpium adversaria nova (A New Notebook of Plants)*. The richness and natural quality of the illustrations carried across the 1,200 plants covered, and for the first time were guided by the principle that all the species were ones Lobelius had himself seen in life.

Here we see print at its best, working to inform the reader. And in Lobelius' second book, printed by the publisher Christopher Plantin, he made an effort to botanically organize the material, which took a step on from Mattioli. Fritillaries rise from the pages, as poised and fresh as in spring today. Plantin built up a huge collection of botanical woodblocks to facilitate the making of new books, and his records show he employed three women full-time as colorists, for which he would charge ten times the price of an uncoloured copy.

BELOW

***Stirpium adversaria nova*, Matthias de l'Obel, second edition, (1576)**

First printed image capturing the graceful *Sarracenia flava*, the carnivorous pitcher plant, here called 'Thuris'.

12 Global botany: Morocco, China, South America and Goa

The increasing urge to classify and document was not just restricted to Europe. Abul Qasim ibn Mohammed al-Ghassani (1548–1610) was a physician at the court of the Saadi Sultanate, who ruled in Morocco in the sixteenth and seventeenth centuries. He is thought to have been of Spanish Moorish descent, which may be why he was picked for a diplomatic mission to the Low Countries. However he is most recognized for being the likely author of *Hadiqat al-azhar fi mahiyyat al-ushb wa-l-aqqar* (*Garden of Flowers in the Explanation of the Character of Herbs and Drugs*).

Spanish Muslim botany had been a significant area of scholarship earlier in the medieval period, with the writings of Ibn al-Baytar (see also page 38). Al-Ghassani's work was the first classification of plants from the Arabic world; elsewhere in the globe this impulse grew also. In China, *Ben Cao* or herbals were increasingly rich compendiums of medicinal knowledge, for example suggesting qinghao or artemisia as an anti-malarial, a use for which it is still considered effective.

In the newly contacted lands of South America, in the 1520s Spain invaded Mexico, establishing a colonial regime in Tenochtitlan, the former capital of the Aztecs. It is clear that the Aztecs possessed a pharmacopeia: a famous garden at Oaxtepec was a repository of herbal medical knowledge for the whole country. Herbals were compiled of

RIGHT

Florentine Codex,
Fray Bernardino de Sahagun,
(*c.* 1545–90)

This huge collaborative work documents Aztec life under the rule of Spain, showing plants in use in many vivid illustrations.

this indigenous knowledge, including the *Cruz-Badiano Codex* of 1552, with recommendations for specific fever treatments, compiled under the aegis of the conquistador Francisco de Mendoza who wanted to export medicinal plants to Spain.

Many Spanish ancillaries came with the conquistadors, including – almost immediately – missionary priests. One such was Fray Bernardino de Sahagún (1499–1590), who was more than usually curious about the language and culture of his new home, and who began preparing a twelve-volume record of the conquered territory, now called New Spain, called *La Historia General de las Cosas de Nueva España* (*General History of the Things of New Spain* or the *Florentine Codex*). For Sahagún this work had both a practical and a divine purpose – the intention was to record the world bilingually, to affirm a two-way communication channel – and he began to learn the language and initiate the translation of the gospels and psalms into Nahuatl, the Aztec language. The workers employed for the making of this work included many native Aztec artists, whose skills reflected their great familiarity with the natural world around them. They would eventually create 2,500 illustrations, with a whole volume devoted to living things, including some spectacular illustrations of plants from within the Aztec tradition.

THIS PAGE

***Florentine Codex*, Fray Bernardino de Sahagun, (*c.*1545–90)**

Evocative illustrations show Aztec musicians playing instruments crafted from plant materials, and growers harvesting corn, a crop that Europe would soon enthusiastically adopt.

This example epitomizes the balance of possible cooperation or coercion that was required in European dealings with indigenous people to extract and record their knowledge. The engravings of Jacques le Moyne's (1533–88) early French explorers in Florida give us a vivid sense of the process of first contact, but one would not suspect from examining them the vivid, edible, lifelike watercolours he made once settled back in Elizabethan London. And Garcia d'Orta (1501–68), a Portuguese physician working in Goa, published the first book on Indian *materia medica* in 1563, *Colóquios dos simples e drogas da Índia* (*Colloquies on the Simples and Drugs of India*), which Carolus Clusius later translated into Latin to make more widely available. D'Orta spoke six or more languages, allowing him a wide range of interactions in the

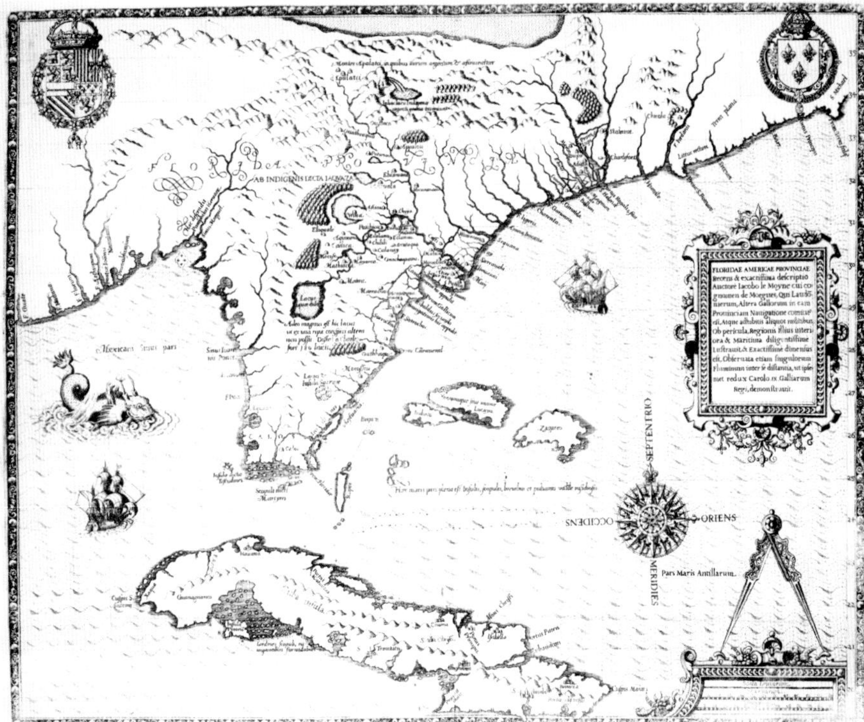

RIGHT
Map of Florida,
Jacques Le Moyne,
(1591)

Published in Frankfurt by Theodor de Bry, this map was made from memory after all the originals were destroyed in an attack by the Spanish.

busy port of Goa, and his book includes Ayurvedic principles as well as the work of Galen and Muslim physician Avicenna.

Increasing numbers of plants and reports arriving from other parts of the world helped to create an enormous thirst for botanical knowledge. This motivated more European nations to seek out foreign plants that they thought might be useful, or that were simply desirable for their beauty. They wanted a share of the botanical riches they saw falling to the preserve of Spain and Portugal alone, with no regard for the sovereignty of indigenous people. It also revealed a problem for the future: how to organize the several thousand plants that botanists had thus far encountered. As the sixteenth century drew to a close, and the world passed into the complexities of early globalized modernity, hints of the scientific method began to emerge. Botanists would soon be using new approaches and tools to study and showcase plants, laying the groundwork to classify the entire flora of the world.

LEFT

Grapes on the vine,
Jacques Le Moyne,
(*c.* 1585)

The Hugenot Le Moyne fled France after massacres in 1572, and settled in London, where he painted remarkable watercolours such as this.

3

BOTANISTS STRIVE TO KNOW AND CLASSIFY MORE PLANTS

(1600–1750)

As the seventeenth century got under way, interest in plants in Europe expanded well beyond viewing them simply through the lens of their medical uses. On the back of spice-seeking Portuguese and Spanish voyages, seeds, bulbs and tubers were starting to arrive regularly in Europe on maritime trade routes from the Americas, Africa and Asia, adding to those plants already travelling overland from the Ottoman Empire. The gardeners, botanists and apothecaries who received these novelties at universities, landowners' estates and physic gardens grew and exchanged them, seeking to answer questions that arose. How did they compare and relate to familiar plants? What kinds of flowers might they have? And how might they serve as a food, medicine or ornament? As they found out more about their exotic charges, they shared their findings in weighty, and increasingly lavish, volumes.

BELOW

Carolus Clusius, (1526–1609)

Clusius cultivated the tulip when it first came to Europe, which helped to establish the Dutch tulip bulb industry.

BOTTOM

Basilius Besler, (1612)

Besler's florilegium *Hortus Eystettensis* is considered to be one of the finest treasures of botanical literature.

1 Plants as objects of garden beauty

Carolus Clusius was among the first scholars in northern Europe to appreciate the beauty, as well as the utility, of plants. Clusius had worked for the Holy Roman Emperor Maximilian II in the imperial gardens in Vienna, before establishing a botanical garden at the University of Leiden in the Dutch Republic (modern-day Netherlands). These roles involved him travelling across Europe in search of specimens, as well as receiving plants from his network of contacts. Ogier Ghiselin de Busbecq sent hyacinths, anemones, crown imperials and tulips, while Philippe de Sivry, Prefect of the city of Mons (in modern-day Belgium) gifted a potato, a plant newly arrived from South America at the hands of the Spanish. Clusius' 1601 work *Rariorum plantarum historia* describes plants he had observed in Spain, Austria and Hungary, new material he had encountered following those trips, and the first treatise on fungi. Meanwhile, his *Exoticorum libri decem* (*Ten Books of Exotica*) of 1605 was dedicated to plants and animals from outside Europe – from the Americas, southeast Asia and Africa. Both volumes were illustrated with woodcuts of plants drawn from botanical specimens.

Clusius had acted as an advisor to botanist and apothecary Basilius Besler (1561–1629), who published the book *Hortus Eystettensis* (*Garden of Eichstätt*) in 1613. Besler worked at the terraced gardens of Willibaldsburg Castle in Bavaria, home of the Prince-Bishop Johann Konrad von Gemmingen of Eichstätt, who commissioned him to have plants in the garden drawn and engraved on copper plates for the publication. Besler's aim was to produce a *florilegium* – a luxurious illustrated work on ornamental flowers – with images as faithful to nature as possible. The finished volume is true to this goal, presenting an array of local and foreign plants by season and at various life stages on 367 plates.

The book shows that spring visitors to Eichstätt might have admired blossoming apple and cherry, along with hyacinths and tulips, while those arriving in summer could have been among the first in Europe to see the showy sunflower, another introduction in the sixteenth century by the Spanish, this time from North America. The extravagant production values and large format of the book, which measured 57 x 46cm (22½ x 18in), were reflected in its price. A coloured copy cost as much as a modest house in Munich at the time. The way that the plants were portrayed, life-sized and detailed to emphasize their beauty, was a great departure from the less-detailed woodcut depictions of plants produced in many herbals.

LEFT

Hortus Eystettensis (Garden of Eichstätt), **Basilius Besler,** (1613)

This book contained engravings of plants growing in the garden of the Willibaldsburg Castle, Eichstätt, Bavaria – such as the lilies and irises depicted below.

RIGHT

Paradisi in Sole Paradisus Terrestris, **John Parkinson, (1629)**

The title page of this book of garden plants depicts Adam and Eve in paradise.

OPPOSITE

John Parkinson, (1657–1650)

Parkinson was both gardener and apothecary.

PARADISI IN SOLE
Paradiſus Terreſtris.
or
A Garden of all ſorts of pleaſant flowers which our
English ayre will permitt to be nourſed vp:
with
A Kitchen garden of all manner of herbes, rootes, & fruites,
for meate or ſauſe vſed with vs,
and
An Orchard of all ſorte of fruitbearing Trees
and ſhrubbes fit for our Land
together
With the right orderinge planting & preſeruing
of them and their vſes & vertues
Collected by John Parkinson
Apothecary of London
1629

Qui veut parangonner l'artifice a Nature
Et nos parcs a l'Eden, indiſcret il meſure.

Le pas de l'elephant par le pas du ciron,
Et de l'Aigle le vol par cil du mouscheron,

The illustrations were produced by copperplate engraving, which gradually replaced woodcut as the preferred method due to its finer details and ability to reproduce more delicate plant structures. This shift in technique allowed for even more precise and intricate botanical illustrations. The technique had become favoured for specialist books about plants from the 1580s onwards; now it rose to dominance. However, the plates wore out easily, which limited print runs and pushed costs even higher. Copperplate engraving thus became the preserve of the enthusiastic amateur and the gardener, while woodcuts were favoured more by botanists, at least for a while.

The English gardener and apothecary John Parkinson (1567–1650) was among the last to use woodcuts when he produced his 1629 volume *Paradisi in Sole Paradisus Terrestris* (*Paradise in the Sun, Paradise on Earth*). Although the work included medicinal plants, it, like Besler's *Hortus*, was more of a florilegium than a herbal, being the first English book about plants to be devoted to their beauty. It contained illustrations of nearly 1,000 plants growing in Parkinson's garden at Long Acre in London, which he had selected for being 'the chiefest for choice, and fairest for shew, from among all the several Tribes and Kindred's of Nature's beauty'. In the book, Parkinson combined his knowledge as a gardener and apothecary to provide a detailed description of each plant along with its origins, followed by information on how best to cultivate it and a short outline of its medical uses. He included flowers, along with herbs and roots, kitchen garden vegetables and orchard fruits. The title page presented a vision of the garden as paradise, with the title itself being a pun: *Paradisi in Sole Paradisus Terrestris* can be translated literally as 'Park-in-Sun's earthly paradise'.

BELOW

***Paradisi in Sole Paradisus Terrestris*, John Parkinson, (1629)**

This engraving depicts camomile, a syrup of which was used to treat jaundice and dropsy (oedema).

Another English florilegium of note was the work of the artist and entomologist Alexander Marshal (1620–82). By the time Marshal was living in South Lambeth, London, in 1641, with the gardener John Tradescant the Younger (1608–62), plants were arriving in Europe from all over the world, including from newly founded French and British settlements in Canada and Virginia, respectively. Tradescant had inherited a garden from his father John Tradescant the Elder (1570–1638) and succeeded him in the royal post of Keeper of the Gardens, Vines and Silkworms at Oatlands Palace in Surrey. The son had himself been to Virginia in search of interesting plants. *Mr Marshal's Flower Book* (1650) included paintings of plants cultivated in various gardens including Tradescant's, and that of Henry Compton, Bishop of London, at Fulham Palace. The images in the book of flowers encompass native plants, such as honeysuckle, and introductions including Spanish love-in-the-mist, the damask rose from the Near East, and a spiderwort from Virginia that had first been grown in England by Tradescant the Elder and now bears the name *Tradescantia virginiana*.

The new genre of flower books helped to raise awareness of the plants being introduced from outside Europe and stimulated demand for them. Before the middle of the sixteenth century, most plants grown in European gardens had been natives of Europe and the Mediterranean, and gardeners craving variety had to be content with double-flowered

BELOW

Paradisi in Sole Paradisus Terrestris, John Parkinson, (1629)

This botanical work was one of the last to use woodcut illustrations. Here, tulips (this page), snake's head fritillaries (opposite page left), and carnations and pinks (opposite page right) are shown.

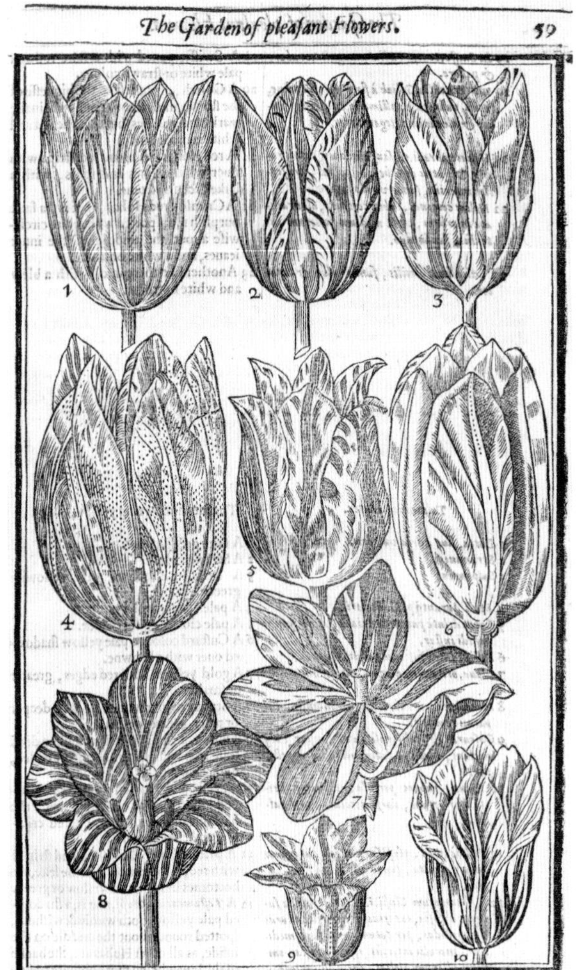

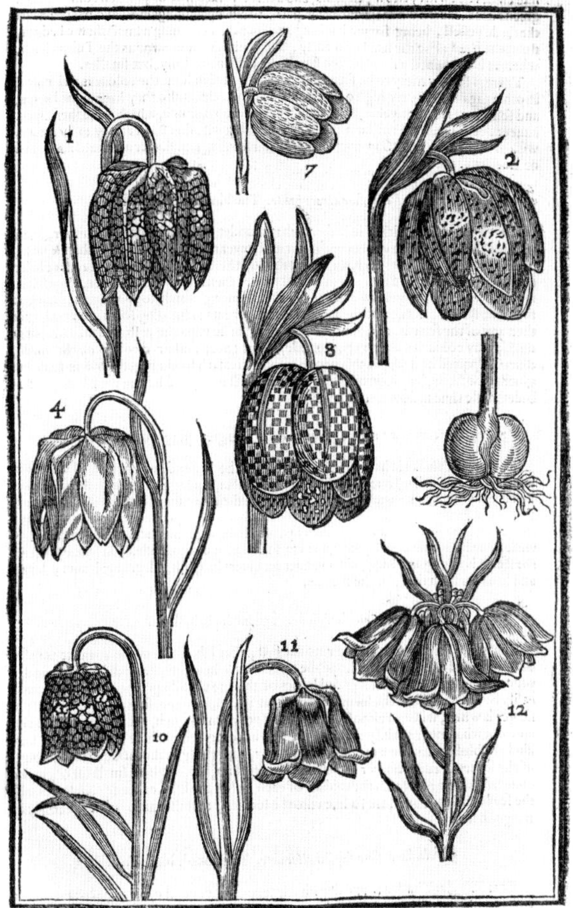
The Garden of pleasant Flowers.
41

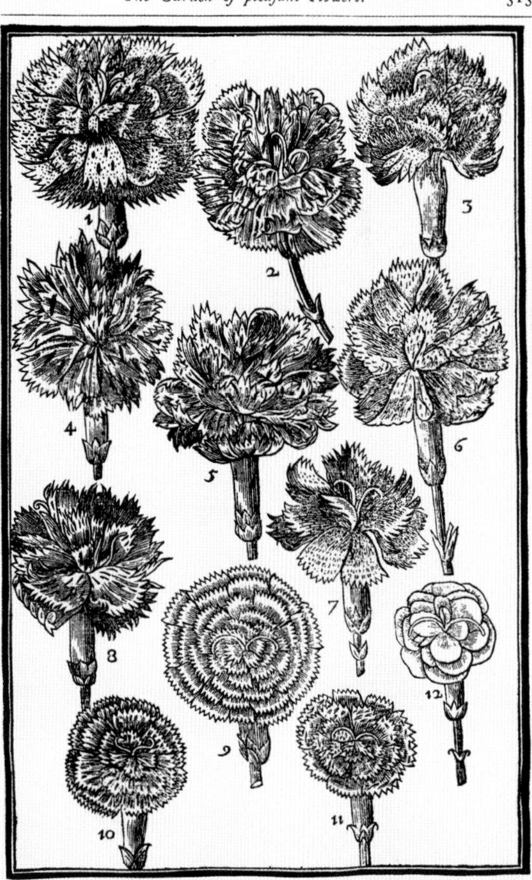
The Garden of pleasant Flowers.
313
1 Heroina Rodolphi florum Imperatoris Princeſſa dictus. Maſter Tuggie his Princeſſe. 2 Caryophyllus Oxonienſis. The French or Oxford Carnation. 3 Caryophyllus Weſtmonaſterienſis. The Gallant or Weſtminſter Gilloflower. 4 Caryophyllus Briſtolienſis. The Briſtow. 5 Caryophyllus Chryſtallinus. The Chryſtall or Chryſtalline. 6 Caryophyllus Sabaudicus ſtriatus. The ſtript Sauadge. 7 Caryophyllus Granatenſis maximus. The Granpere or greateſt Granado. 8 Caryophyllus peramanus. The Dainty. 9 Caryophyllus Silcſiacus maximus Ingonij Ioannis. Iohn Witty his great tawny Gilloflower. 10 Caryophyllus Sileſiacus ſtriatus. The ſtript Tawny. 11 Caryophyllus marmor-amulus. The marbled Tawny. 12 Caryophyllus roſeus rotundus magiſtri Tuggie. Maſter Tuggie his Roſe Gilloflower.

ABOVE

John Tradescant the Younger, (1608–62)

A botanist, gardener, traveller and collector, Tradescant was a contemporary of artist Alexander Marshal.

RIGHT

***Mr Marshal's Flower Book*, Alexander Marshal, (1650–82)**

English gardener and artist Alexander Marshal painted this image of sunflower with greyhound soon after the sunflower had arrived in England from North America for the first time.

LEFT

Mr Marshal's Flower Book, **Alexander Marshal, (1650–82)**

'Jay with crocuses and anemones'. Marshal painted, not as a professional, but to share images with friends and fellow gardeners.

RIGHT

Mr Marshal's Flower Book,
Alexander Marshal,
(1650)

Crown imperial, narcissi and auriculas (this page) and hyacinths, Persian iris, Spanish daffodils and purple crocus (opposite page). Marshal captured plants' beauty in exquisite detail (overleaf).

forms, deformities and multiple colours. A hundred years on and crocuses, leucojums, erythroniums, ornithogalums, cyclamens, hyacinths, lilies, fritillaries, ranunculus and tulips from the Ottoman Empire were springing up in Europe's flowerbeds. Of these, tulips were particularly beguiling to gardeners because of their colours and diversity. In 1601, Carolus Clusius had devoted fifteen pages to tulips in *Rariorum plantarum historia*, while Besler's *Hortus* presented twelve plates of mainly goblet or star-shaped kinds, mostly in shades of yellow, white and red. By 1629, John Parkinson was able to list more than one hundred varieties in *Paradisi in Sole Paradisus Terrestris*, describing characteristics such as those of 'Crimson Fooles Coat, a dark crimson, and pale white empaled together' and 'Fooles Cappe, that is, with lists of stripes of yellow running through the middle of every leafe of the red.'

BELOW

***Rariorum plantarum historia*, Carolus Clusius, (1601)**

By the start of the 17th century, tulips were beginning to pique the curiosity of botanists in Europe.

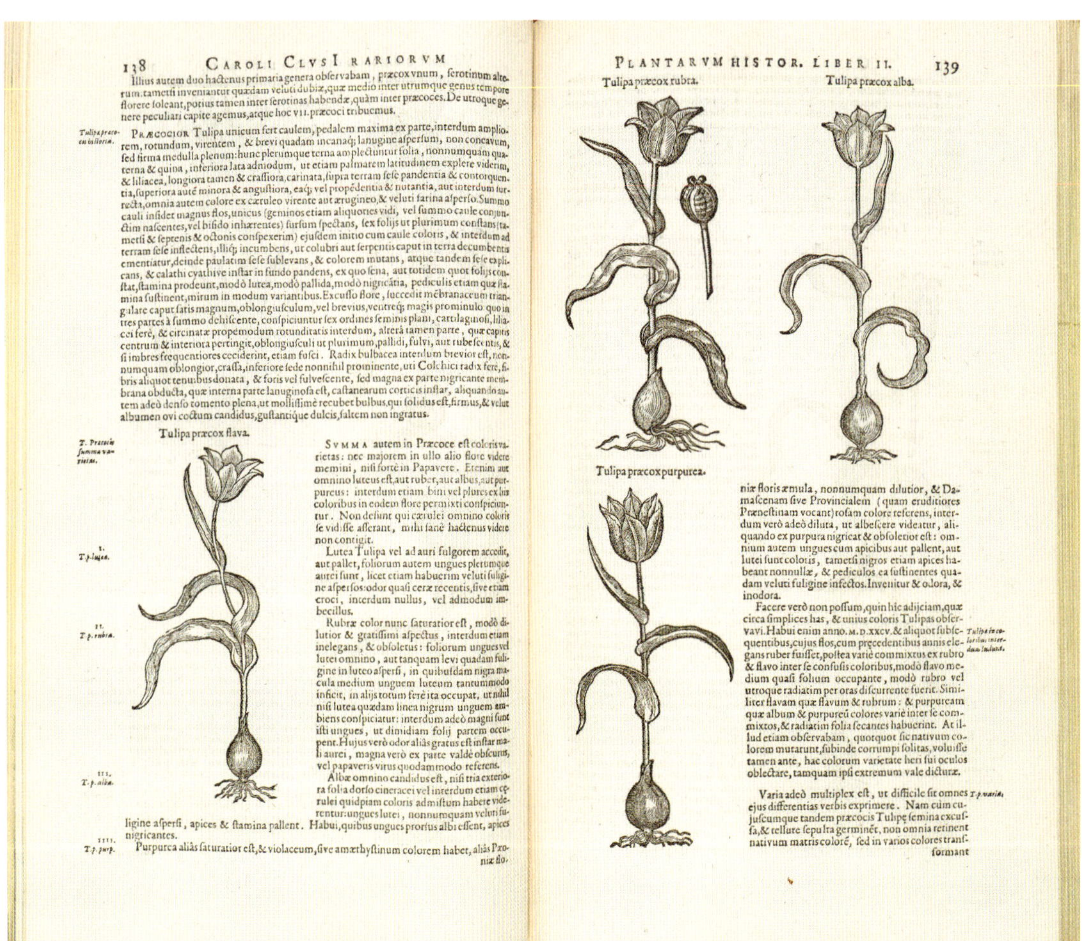

138 CAROLI CLVSI RARIORVM

Illius autem duo hactenus primaria genera observabam, præcox vnum, serotinum alterum: tametsi inveniantur quædam veluti dubiæ, quæ medio inter utrumque genus tempore florere soleant, potius tamen inter serotinas habendæ, quàm inter præcoces. De utroque genere peculiari capite agemus, atque hoc VII. præcoci tribuemus.

Tulipa præcocis historia. PRÆCOCIOR Tulipa unicum fert caulem, pedalem maxima ex parte, interdum ampliorem, rotundum, virentem, & brevi quadam incanaq; lanugine aspersum, non concavum, sed firma medulla plenum: hunc plerumque terna amplectuntur folia, nonnumquam quaterna & quina, inferiora lata admodum, ut etiam palmarem latitudinem explere viderim, & liliacea, longiora tamen & crassiora, carinata, supra terram sese pandentia & contorquentia, superiora auté minora & angustiora, eaq; vel propédentia & nutantia, aut interdum surrecta, omnia autem colore ex cæruleo virente aut ærugineo, & veluti farina asperso. Summo cauli insidet magnus flos, unicus (geminos etiam aliquoties vidi, vel summo caule conjunctim nascentes, vel bifido inhærentes) sursum spectans, sex folijs ut plurimum constans (tametsi & septenis & octonis conspexerim) ejusdem initio cum caule coloris, & interdum ad terram sese inflectens, illiq; incumbens, ut colubri aut serpentis caput in terra decumbentis ementiatur, deinde paulatim sese sublevans, & colorem mutans, atque tandem sese explicans, & calathi cyathive instar in fundo pandens, ex quo sena, aut totidem quot folijs constat, stamina prodeunt, modò lutea, modò pallida, modò nigricátia, pediculis etiam quæ stamina sustinent, mirum in modum variantibus. Excusso flore, succedit mēbranaceum triangulare caput satis magnum, oblongiusculum, vel brevius, ventreq; magis prominulo: quo in tres partes à summo dehiscente, conspiciuntur sex ordines seminis plani, cartilaginosi, liliacei ferè, & circinatæ propemodum rotunditatis interdum, alterá tamen parte, quæ capitis centrum & interiora pertingit, oblongiusculi ut plurimum, pallidi, fulvi, aut rubescentis, & si imbres frequentiores ceciderint, etiam fusci. Radix bulbacea interdum brevior est, nonnumquam oblongior, crassa, inferiore sede nonnihil prominente, uti Colchici radix ferè, fibris aliquot tenuibus donata, & foris vel fulvescente, sed magna ex parte nigricante membrana obducta, quæ interna parte lanuginosa est, castanearum corticis instar, aliquando autem adeò denso tomento plena, ut mollissimè recubet bulbus, qui solidus est, firmus, & velut albumen ovi coctum candidus, gustantique dulcis, saltem non ingratus.

Tulipa præcox flava.

T. Præcocis summa varietas. SVMMA autem in Præcoce est coloris varietas: nec majorem in ullo alio flore videre memini, nisi fortè in Papavere. Etenim aut omnino luteus est, aut ruber, aut albus, aut purpureus: interdum etiam bini vel plures ex his coloribus in eodem flore permixti conspiciuntur. Non desunt qui cærulei omnino coloris se vidisse asserant, mihi sanè hactenus videre non contigit.

I. T. p. lutea. Lutea Tulipa vel ad auri fulgorem accedit, aut pallet, foliorum autem ungues plerumque aurei sunt, licet etiam habuerim veluti fuligine aspersos: odor quasi ceræ recentis, sive etiam croci, interdum nullus, vel admodum imbecillus.

II. T. p. rubra. Rubræ color nunc saturatior est, modò dilutior & gratissimi aspectus, interdum etiam inelegans, & obsoletus: foliorum ungues vel lutei omnino, aut tanquam levi quadam fuligine in luteo aspersi, in quibusdam nigra macula medium unguem luteum tantummodo inficit, in alijs totum ferè ita occupat, ut nihil nisi lutea quædam linea nigrum unguem ambiens conspiciatur: interdum adeò magni sunt isti ungues, ut dimidiam folij partem occupent. Hujus verò odor aliâs gratus est instar mali aurei, magna verò ex parte valdè obscurus, vel papaveris virus quodammodo referens.

III. T. p. alba. Albæ omnino candidus est, nisi tria exteriora folia dorso ciceracei vel interdum etiam cærulei quidpiam coloris admistum habere viderentur: ungues lutei, nonnumquam veluti fuligine aspersi, apices & stamina pallent. Habui, quibus ungues prorsus albi essent, apices nigricantes.

IIII. T. p. purp. Purpurea aliâs saturatior est, & violaceum, sive amæthystinum colorem habet, aliâs Pronia flo-

PLANTARVM HISTOR. LIBER II. 139

Tulipa præcox rubra.

Tulipa præcox alba.

Tulipa præcox purpurea.

niæ floris æmula, nonnumquam dilutior, & Damascenam sive Provincialem (quam eruditiores Prænestinam vocant) rosam colore referens, interdum verò adeò diluta, ut albescere videatur, aliquando ex purpura nigricat & obsoletior est: omnium autem ungues cum apicibus aut pallent, aut lutei sunt coloris, tametsi nigros etiam apices habeant nonnullæ, & pediculos ea sustinentes quadam veluti fuligine infectos. Invenitur & odora, & inodora.

Facere verò non possum, quin hic adjiciam, quæ circa simplices has, & unius coloris Tulipas observavi. Habui enim anno M.D.XXCV. & aliquot subsequentibus, cujus flos, cum præcedentibus annis elegans ruber fuisset, postea variè commixtus ex rubro & flavo inter se confusis coloribus, modò flavo medium quasi folium occupante, modò rubro vel utroque radiatim per oras discurrente fuerit. Similiter flavam quæ flavum & rubrum: & purpuream quæ album & purpureū colores variè inter se commixtos, & radiatim folia secantes habuerint. At illud etiam observabam, quotquot sic nativum colorem mutarunt, subinde corrumpi solitas, voluisse tamen ante, hac colorum varietate heri sui oculos oblectare, tamquam ipsi extremum vale dicturæ.

Varia adeò multiplex est, ut difficile sit omnes ejus differentias verbis exprimere. Nam cùm cujuscumque tandem præcocis Tulipę semina excussa, & tellure sepulta germinēt, non omnia retinent nativum matris colorē, sed in varios colores transformant T. p. varia.

Tulips frequently produced new colour patterns, with previously single-colour bulbs exhibiting vibrant stripes, flames and feathers of colour. We now know that a disease called tulip breaking virus was responsible for this unpredictable phenomenon but the cause was still unknown in the early seventeenth century. When new flamboyantly patterned variations appeared spontaneously, growers sought to reproduce them by vegetative propagation, prompting the first large-scale commercial exploitation of a garden plant. By the 1620s, the prices of the most spectacular varieties were rapidly rising, with the striking red-and-white striped 'Semper Augustus' variety selling for 1,000 florins a bulb (around US$60,000 in today's money). Speculation pushed prices even higher; within a decade, one 'Semper Augustus' bulb had exchanged hands for 10,000 guilders (US$600,000). Then, in 1637, the market was flooded and crashed. Previously highly priced bulbs became worthless overnight.

While the crisis caused by 'tulipomania' was not repeated for other plants, the desire for other novelties helped to nurture trade in plants. In London, although plant and seed sellers had existed from the early sixteenth century, business had tended to be conducted locally. By John Tradescant the Elder's day, wealthy estate owners wanting new plants for their grounds would either exchange them with other landowners or source them from nurseries across Europe. An intriguing publication dating from this time is *Tradescant's Orchard*, a leather-bound volume containing sixty-six images of fruit varieties, currently held in the Bodleian Library in Oxford, UK. Ranging from 'the Dwarfe cherry' to the 'grete Early yellowe peech' and the 'grete Roman Hasell Nut', the paintings probably represent the earliest surviving display of painted fruits growing in England. They were not produced by Tradescant but he is referenced in the book as 'JT'.

ABOVE
***Tradescant's Orchard*, unknown, (early 17th century)**

This collection of paintings of fruit varieties may have been an early horticultural catalogue. This rare and unique record is held at the Bodleian Library, Oxford, UK.

It is possible that the collection of images was a form of nursery catalogue. The earliest such list known from Europe is *Florilegium amplissimum et selectissimum* (*A Most Extensive and Select Collection of Flowers)* from 1612, showcasing 560 unusual garden plants, including many bulbs, that were the wares of the Dutch painter and trader Emanuel Sweert. However, the next illustrated flower and seed catalogue known from England, nurseryman Robert Furber's *Twelve Months of Flowers*, was not published until 1730. A commercial venture between Furber, the Dutch artist Pieter Casteels II and the English engraver Henry Fletcher, it comprises twelve beautifully illustrated plates showcasing more than 400 flowers. They included auriculas, anemones, roses, tulips and hyacinths, aimed at the increasing number of people growing flowers for pleasure.

2 Botanists catalogue plants in geographic 'flora'

While gardeners to the rich and famous were showcasing their latest blooms in beauty-focused florilegia, botanists were publishing their attempts to classify plants in another new kind of publication, now known as a flora. These books loosely sought to list and

RIGHT AND OVERLEAF

Twelve months of flowers, **Robert Furber, (1730)**

These hand-coloured engravings depicting flowers of July (this page), September (opposite page), August (overleaf left) and October (overleaf right) were made after paintings by Dutch artist Pieter Casteels II.

1 Red Sow Bread.
2 White Sow Bread.
3 White Corn-marigold.
4 New Tree Primrose.
5 Four leav'd Geranium.
6 Quill'd African Marigold.
7 Hearts ease.
8 Shrub Cotton.
9 Sheffords Hester Auricula.
10 Virginian Birthwort.
11 Virginian upright Bramble.
12 Scarlet Indian Cane.
13 White Colchicum.
14 Bean Caper.
15 All red Amaranthus.
16 Double white Sapwort.
17 Yellow Indian Cane.
18 Virginian Poke.
SEPTEMBER
19 Gentianella.
20 White monthly Rose.
21 Yellow Amaranthus.
22 Oriental Arssmart.
23 Broad leav'd Cardinal.
24 Yellow Colchicum.
25 Hardy golden Rod.
26 White Althaea frutex.
27 Chequer'd Colchicum.
28 Yellow Colutea.
29 Dwarf Pomgranate.
30 Strip'd single Female balsom.
31 African Marigold.
32 Honour & glory Auricula.
33 White flower Moth Mullein.
34 Double Colchicum.
35 Three leav'd Passion flower.
Design'd by Pet. Casteels
From the Collection of Robt. Furber Gardiner at Kensington 1730
Engrav'd by H. Fletcher

1 Purple Althæa frutex.
2 Ivy leav'd Jasmine.
3 Iris Uvaria.
4 Purple Sultan.
5 Purple toad flax.
6 Purple Amaranthoides.
7 Double Arabian Jasmine.
8 Yellow Ketmia.
9 Purple Coxcomb Amaranth.
10 Shrub St. Johns wort.
11 Pona's blew Throatwort.
12 Palma Christi.
13 Purple Convolvulus.
14 Polyanthos.
15 Indian yellow Jasmine.
16 Double flowering Myrtle.
AUGUST
17 Egyptian scarlet holly hock.
18 Yellow strip'd marvel of peru.
19 Strip'd Monthly rose.
20 Double fether few.
21 Semper Augustus Auricula.
22 Dwarf Convolvulus.
23 Willow leav'd Apocynum.
24 Apios of America.
25 Virginian flowering Raspberry.
26 Lisols from Genoa.
27 Double spanish Jasmine.
28 White Eternal.
29 Fruit bearing Passion flower.
30 Scarlet Althæa.
31 Canary shrub fox glove.
32 Long blowing honey suckle.
33 Double purple Virgins bower.
34 Virginian scarlet Martagon.

1 Tuberose flower.
2 Single Nasturtium.
3 Yellow peren.t Poppy.
4 Purple Polyanthos.
5 Saffron flower.
6 Strip'd double Colchicum.
7 Single blew Periwinkle.
8 Trumpet flower.
9 Camomile double.
10 Semper Augustus Auricula.
11 Indian Tobacco.
12 Arbutus double.
13 Best flowering Geranium.
14 Guernsey Lilly.
15 Autumn Carnation.
16 Agnus Castus.
OCTOBER
17 Long blowing Honeysuckle.
18 Spiked Aster.
19 Belladona Lilly.
20 Ever green Honeysuckle.
21 Leonurus, or Archangel tree.
22 Black Granes bill.
23 Scarlet Granes bill.
24 Marigold tree.
25 Musk Scabious.
26 Double white Musk rose.
27 Box leav'd Myrtle.
28 Michaelmas Daisie.
29 Yellow Passion flower.
30 Hollyhock always double.
31 Virgina Shrub Acre.
Engrav'd by H. Fletcher

PRODROMI

Gramen caninum maritimum asperum.

Gramen caninum maritimum spicatum.

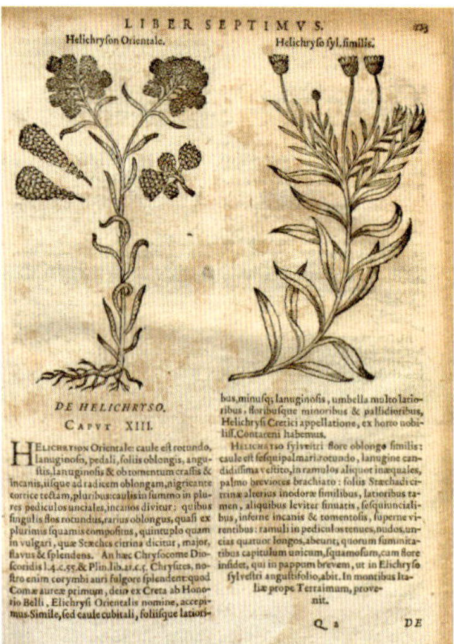

LIBER SEPTIMVS.

Helichryson Orientale.

Helichryso syl. similis.

DE HELICHRYSO.

CAPVT XIII.

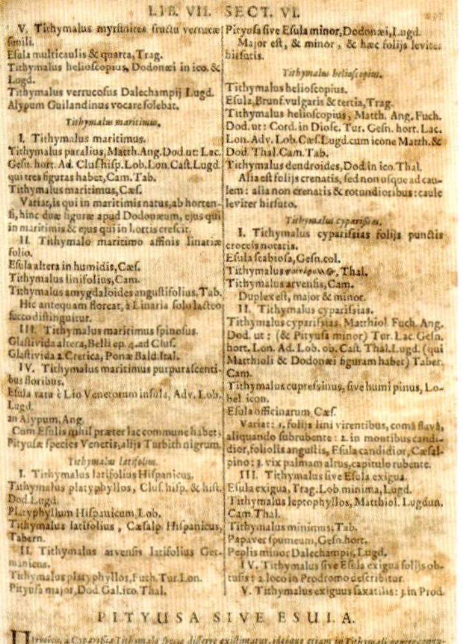

LIB. VII. SECT. VI.

PITYUSA SIVE ESULA.

describe plants that occurred in one place, be that a small area, region, country or the world. An early local work, *Sylva harcynia* (*Forests of the Harz region*), published posthumously in 1588 by Johannes Thal (1542–83), had helped to establish conventions for this new kind of botanical tome. Covering the plants of the Harz mountains in Germany, it gave localities, alphabetically listed vernacular names and had descriptions of each entry's scientific and economic characteristics. From 1600, works of this nature began to proliferate.

Caspar Bauhin's 1623 *Pinax Theatri Botanici* (*Illustrated Exposition of Plants*), covering 6,000 plant types, could be considered as the first true 'world list' of plants. Synonyms that Bauhin used were later incorporated in John Parkinson's 1640 compilation of 3,800 medicinal plants, *Theatrum botanicum* (*Botanical Theatre*), which he considered to represent the plants known from all the four continents of the world, including twenty-eight new British species. Meanwhile *Flora Danica* (*Danish Flora*), published in 1648 by the Danish physician and naturalist Simon Paulli (1603–80) of the University of Copenhagen, was an early attempt to present a national flora. The book, which contained 384 seasonally arranged woodcut illustrations and descriptions of wild Danish and Norwegian plants of medicinal interest, helped to establish the term flora for place-based botanical works.

In England, a decade later, John Ray (1627–1705) was working on a similar project to *Flora Danica*. Between 1658 and 1662, he made four journeys exploring the Midlands and west coasts of England and Wales, and the east coast of England and parts of Scotland, seeking out and describing first-hand the plants he encountered. His write-ups of his travels provide interesting insights into the natural habitats that existed in the seventeenth century, many of which have long gone, as well as the social context of the time. They also include the early use of simple Latin names for plants. In Cornwall he observed that 'nothing [is] more common than *Osmunda regalis* about springs and

72 Den Anden Part

II. Laurus vulgaris. *Bauhin.*
* Laurus. *Brunf. Trag. Dodon. Lon.Lob.*
* Laurus tenuifolia. *Matth.*
* Laurus altera ſpecies. *Dod.*

60. Laurus.

61. Leu-

OPPOSITE TOP

Pinax Theatri Botanici (Illustrated Exposition of Plants), Caspar Bauhin, (1623)

This work was an early attempt to systematically describe and classify plants.

OPPOSITE BELOW

John Ray, (1627–1705)

John Ray was a blacksmith's son who became regarded for classifying plants he observed on his travels across Britain and Continental Europe.

THIS PAGE

Flora danica, Simon Paulli, (1648)

Paulli's work was an early attempt to classify plants within a particular geographic area.

RIGHT

Catalogus plantarum Angliae,
John Ray,
(Second edition, 1677)

John Ray made detailed observations of plants he encountered on extensive travels to inform this work. The plant pictured here is recognisable as the yellow-flowered shrubby cinquefoil, now named as *Potentilla fruticosa*.

rivulets in this country'; populations of this plant were later decimated by Victorian fern collectors. After travelling by 'a long and bad way' to Bala and Dolgellau (Dolgelly), Wales, Ray delighted at seeing the globeflower (*Trollius europaeus*) for the first time, an event that took place shortly before he learned of the death of Oliver Cromwell, England's de facto ruler at the time. Meanwhile, in Scotland, Ray reported encounters with *Silene maritima*, *Lavatera arborea* and *Cochlearia officinalis*, as well as being in the country at a time when 'diverse women were burnt for witches'. He published *Catalogus plantarum circa Cantabrigiam* (*Catalogue of Cambridge Plants*) in 1660 and *Catalogus plantarum Angliae* (*Catalogue of English Plants*) in 1670.

A conundrum for European botanists was how to work out if a plant described in one flora was the same as one presented in another, and this was becoming increasingly urgent as more and more plants arrived on the continent. As far back as 1536, the botanist Antonio Musa Brassavola (1500–55) had realized that: 'not a hundredth part of the herbs existing in the whole world was described by Dioscorides, not a hundredth part by Theophrastus or Pliny, but we add more every day . . .' Over time, as botanists had encountered ever wider ranges of plants, their approaches as to how best to classify nature

BELOW

***Catalogus plantarum Angliae*,**
John Ray,
(1670)

Title page and an image from this work. Today, Ray's images and descriptions of plants tell us about the past biodiversity of the places he visited.

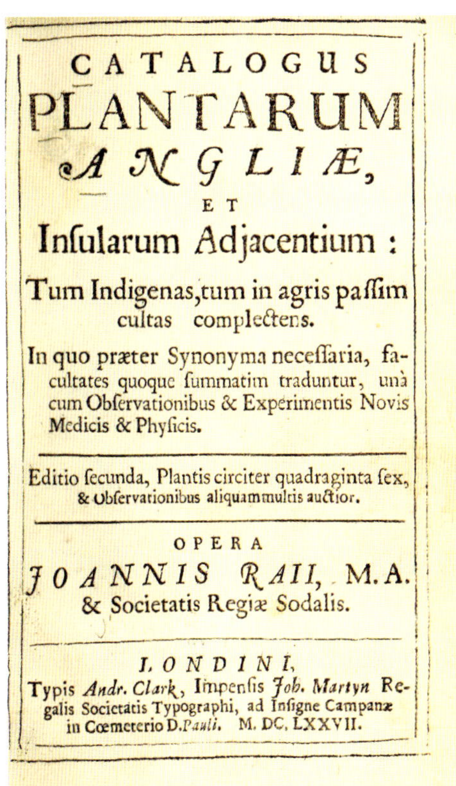

CATALOGUS
PLANTARUM
ANGLIÆ,
ET
Inſularum Adjacentium :
Tum Indigenas, tum in agris paſſim cultas complectens.
In quo præter Synonyma neceſſaria, facultates quoque ſummatim traduntur, unà cum Obſervationibus & Experimentis Novis Medicis & Phyſicis.

Editio ſecunda, Plantis circiter quadraginta ſex, & Obſervationibus aliquammultis auctior.

OPERA
JOANNIS RAII, M.A.
& Societatis Regiæ Sodalis.

LONDINI,
Typis *Andr. Clark*, Impenſis *Joh. Martyn* Regalis Societatis Typographi, ad Inſigne Campanæ in Cœmeterio D.*Pauli*. M. DC. LXXVII.

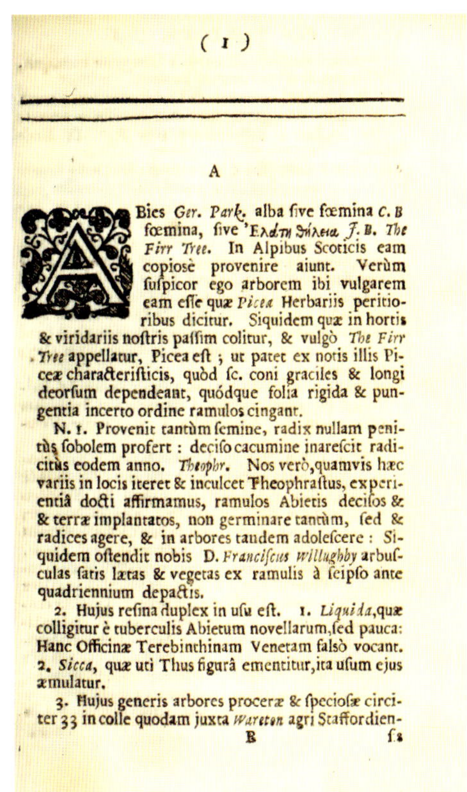

(1)

A

ABies *Ger. Park.* alba ſive fœmina *C. B* fœmina, ſive 'Ελάτη θήλεια *J. B. The Firr Tree.* In Alpibus Scoticis eam copioſè provenire aiunt. Verùm ſuſpicor ego arborem ibi vulgarem eam eſſe quæ *Picea* Herbariis peritioribus dicitur. Siquidem quæ in hortis & viridariis noſtris paſſim colitur, & vulgò *The Firr Tree* appellatur, Picea eſt ; ut patet ex notis illis Piceæ characteriſticis, quòd ſc. coni graciles & longi deorſum dependeant, quódque folia rigida & pungentia incerto ordine ramulos cingant.

N. 1. Provenit tantùm ſemine, radix nullam penitùs ſobolem profert : deciſo cacumine inareſcit radicitùs eodem anno. *Theophr.* Nos verò, quamvis hæc variis in locis iteret & inculcet Theophraſtus, experientiâ docti affirmamus, ramulos Abietis deciſos & & terræ implantatos, non germinare tantùm, ſed & radices agere, & in arbores tandem adoleſcere : Siquidem oſtendit nobis D. *Franciſcus Willughby* arbuſculas ſatis lætas & vegetas ex ramulis à ſeipſo ante quadriennium depactis.

2. Hujus reſina duplex in uſu eſt. 1. *Liquida*, quæ colligitur è tuberculis Abietum novellarum, ſed pauca: Hanc Officinæ Terebinthinam Venetam falsò vocant. 2. *Sicca*, quæ uti Thus figurâ ementitur, ita uſum ejus æmulatur.

3. Hujus generis arbores proceræ & ſpecioſæ circiter 33 in colle quodam juxta *Wareton* agri Staffordien-

B ſi

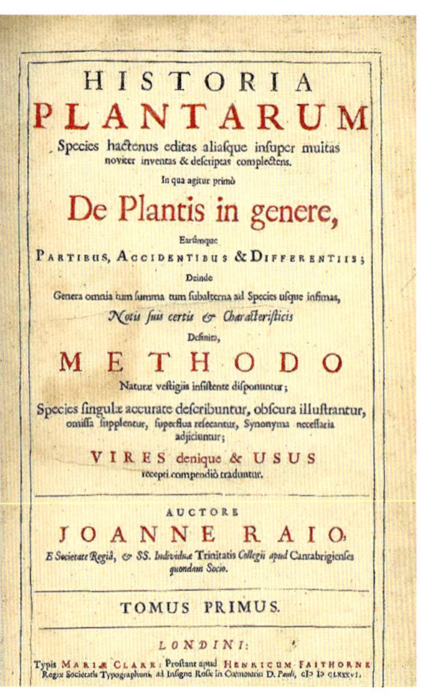

HISTORIA
PLANTARUM
Species hactenus editas aliasque insuper multas noviter inventas & descriptas complectens.
In qua agitur primò
De Plantis in genere,
Earúmque
PARTIBUS, ACCIDENTIBUS & DIFFERENTIIS;
Deinde
Genera omnia tum summa tum subalterna ad Species usque infimas,
Notis suis certis & Characteristicis
Definita,
METHODO
Naturæ vestigiis insistente disponuntur;
Species singulæ accurate describuntur, obscura illustrantur, omissa supplentur, superflua resecantur, Synonyma necessaria adjiciuntur;
VIRES denique & USUS
recepti compendiò traduntur.

AUCTORE
JOANNE RAIO,
E Societate Regiâ, & SS. Individuæ Trinitatis *Collegii apud* Cantabrigienses *quondam Socio.*

TOMUS PRIMUS.

LONDINI:
Typis MARIÆ CLARK: Prostant apud HENRICUM FAITHORNE Regiæ Societatis Typographum, ad Insigne Rosæ in Cœmeterio *D. Pauli*, CIↃ IↃ CLXXXVI.

ABOVE

Historia plantarum,
John Ray,
(1686–1704)

Seeking to systematically classify plants, John Ray was the first to define the word 'species'.

RIGHT

Élémens de Botanique ou Métode pour Connoître les Plantes (*Elements of Botany or Methods for Knowing Plants*),
Joseph Pitton de Tournefort,
(1694)

This work classified all the plants known in the Western world at the time, covering 10,146 species in 648 genera.

ÉLEMENS
DE
BOTANIQUE:
OU
MÉTODE
POUR CONNOÎTRE LES PLANTES.

IDÉE GENERALE DE LA BOTANIQUE, avec une Histoire abregée de cette Sience.

LA Botanique ou la Sience qui traite des Plantes, a deux parties qu'il faut distinguer avec soin : La connoissance des plantes, & celle de leurs vertus.

Connoître les plantes, c'est précisément savoir les noms qu'on leur a donné par raport à la structure de quelques-unes de leurs parties. Cette structure fait le caractere qui distingue essentiellement les plantes les unes d'avec les autres. L'idée de ce caractere

A

had evolved. In ancient Greece, Theophrastus (see also page 20–23) had classified plants on the basis of their form (herbs, shrubs or trees), their lifespan (annual, biennial or perennial), as well as floral characteristics. During the thirteenth century, Albert the Great (see also page 43) had employed the difference between monocotyledons (plants producing a single first leaf from each seed) and dicotyledons (plants with paired seed leaves), still today one of the most significant divisions in the plant kingdom. In 1583, the Italian physician Andrea Cesalpino had divided plants into herbaceous and woody kinds, and realized the value of using features within flowers and fruits, rather than form or habit, to classify plants. And in his *Pinax* and other works, Caspar Bauhin (see also pages 106–107) applied a relatively simple naming system that is echoed by the two-word Latin names used today. But an entirely satisfactory system remained elusive.

It was John Ray who made the first real major advance in taxonomy (the branch of science concerned with classification). Concerned by the 'multiplication of species' that could arise if one plant was classified differently by two people, Ray combined the approaches of several earlier botanists. He first drew on Theophrastus, dividing plants into shrubs, herbs and trees. Then he split these into fifteen classes of dicotyledons and four classes of monocotyledons. Some of his groups equate to present-day families, such as Brassicaceae (Tetrapetalae), Lamiaceae (Verticillatae) and grasses (Staminae). In two of Ray's subsequent works, *Methodus plantarum nova* (*A New Method of Planting*, 1682) and *Historia plantarum* (*History of Plants*, 1686–1704) he urged his fellow botanists to consider all parts of plants during classification attempts. Ray coined words that are still used today, such as petal and pollen. He also provided the first biological definition of the term species, explaining it as a group of individuals that could potentially reproduce with each other.

Ray's contemporary, Pierre Magnol (1638–1715), who lived and worked in Montpellier, France, honed this classification by grouping plants into seventy-six families in his 1689 book *Prodromus historiae generalis, in qua familiae per tabulas disponuntur* (*Prologue to the General History, In Which the Families are Arranged by Tables*). In doing so, he established the 'family' concept still used today. And Magnol's student Joseph Pitton de Tournefort (1656–1708), who became professor of botany at Jardin de Roi, Montpellier, France, contributed further to classification in his clearly written and illustrated work of 1694 *Élémens de botanique ou méthode pour connaître les plantes* (*Elements of Botany or Methods for Knowing Plants*).

Tournefort arranged plants with petals into classes based on the form of the corolla; into families according to where the corolla was positioned; and into genera on the basis of fruit and seed characteristics. He is considered to have devised the concept of the 'genus', the useful grouping of very closely related species that remains today the first part of any organism's Latin name. Although this classification was entirely unnatural because it used just a few characters to classify all the included species, and, in some ways, was a retrograde step from Ray's system, it was widely used for the next fifty years.

RIGHT

Nova Plantarum, animalium et mineralium Mexicanorum historia, Antonio Recchi, (1651)

This condensed version of Francisco Hernández's work was published 80 years after the naturalist began recording the natural history of Mexico.

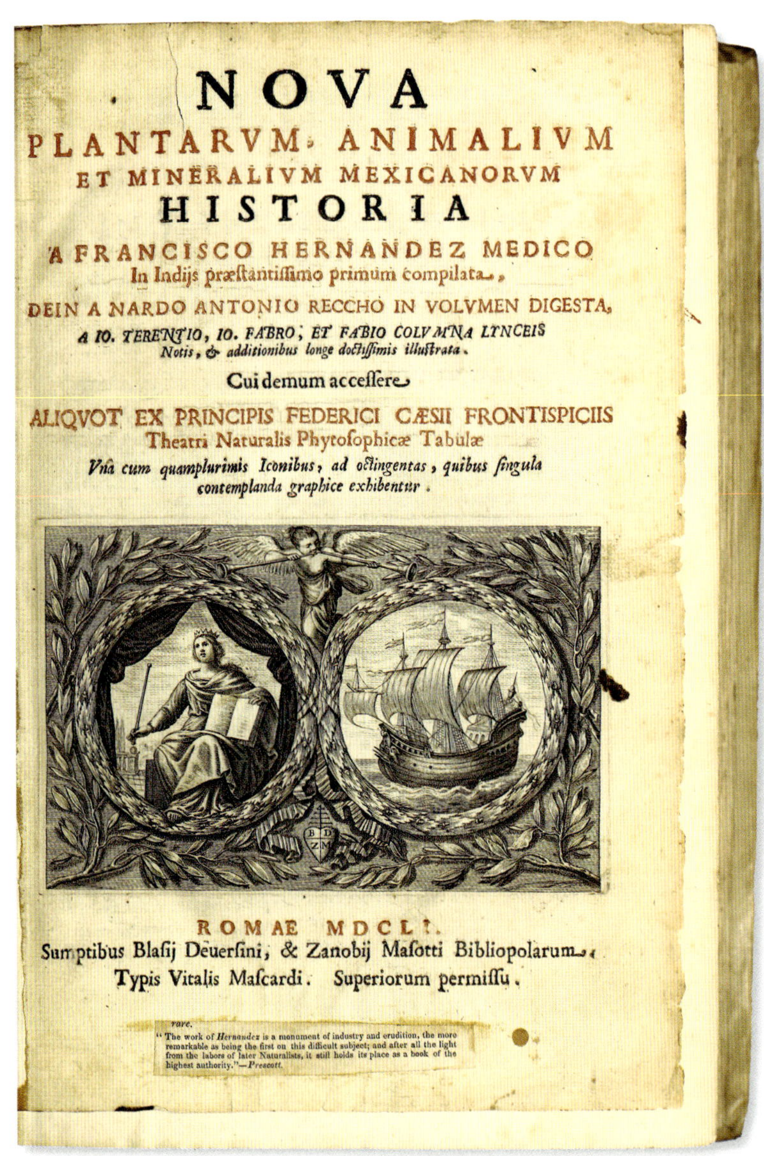

NOVA
PLANTARVM, ANIMALIVM
ET MINERALIVM MEXICANORVM
HISTORIA
A FRANCISCO HERNANDEZ MEDICO
In Indijs præstantissimo primum compilata,
DEIN A NARDO ANTONIO RECCHO IN VOLVMEN DIGESTA,
A IO. TERENTIO, IO. FABRO, ET FABIO COLVMNA LYNCEIS
Notis, & additionibus longe doctissimis illustrata.
Cui demum accessere
ALIQVOT EX PRINCIPIS FEDERICI CÆSII FRONTISPICIIS
Theatri Naturalis Phytosophicæ Tabulæ
Vna cum quamplurimis Iconibus, ad octingentas, quibus singula contemplanda graphice exhibentur.

ROMAE MDCLI.
Sumptibus Blasij Deuersini, & Zanobij Masotti Bibliopolarum,
Typis Vitalis Mascardi. Superiorum permissu.

rare.
"The work of *Hernandez* is a monument of industry and erudition, the more remarkable as being the first on this difficult subject; and after all the light from the labors of later Naturalists, it still holds its place as a book of the highest authority."—*Prescott.*

3 Europe's imperialism drives interest in 'exotics'

Western botany was advancing at this time within the context of aggressive colonialism. During the fifteenth and sixteenth centuries, the Spanish and Portuguese had located, explored, settled and exploited resources within the Americas, Africa and Asia, motivated in part by the desire to secure supplies of plant-based commodities, such as pepper, cloves and nutmeg. In the seventeenth century, they continued with their expansionist mission, but now England, France and the Netherlands saw the possible benefits and launched themselves into efforts to catch up, also becoming significant colonizing forces.

As these countries vied to establish trading posts, mercantile networks and colonies abroad, they ran roughshod over the indigenous peoples they encountered and any claims they might have to their own botanical resources. Not only did the Europeans introduce diseases that decimated local communities, they then imposed their own cultural, religious and political systems, causing conflicts, displacement and cultural assimilation. Meanwhile, in moving plants from one place to another they affected natural environments, for example by introducing 'invasive' species that went on to run riot in their new terrains.

Mercantilism, a form of economic nationalism in which European countries competed with each other to amass wealth by exporting more goods than they imported, drove exploitation in the foreign nations they entered, and led to conflicts among the colonial aggressors, too. In the Americas, plantation economies using enslaved labourers to drive large-scale agricultural production rose to prominence. In Asia, trade became dominated by companies such as the Dutch East India Company (*Vereenigde Oost-Indische Compagnie*, VOC) and English (later British) East India Company (EIC). Although these companies' primary purpose was buying and selling goods, in time they became colonial powers themselves, making treaties, building fortifications and carrying out military operations. With plants being the primary providers of the food, medicine and useful materials that the European nations sought to trade, botany lay at the heart of the exploitative colonial mercantile system. Herbals, flora and other botanical works, often compiled using indigenous knowledge, were the vehicle through which botanical understanding was acquired, shared and used to expand European economies.

Rerum medicarum Novae Hispaniae thesaurus (*Treasury of the Medicinal Things of New Spain*) by Francisco Hernández (1517–1587) covering modern-day Mexico provides an example (see also page 52). Compiled at the request of King Philip II of Spain, it contains descriptions of 3,000 plants, which were named through interviews with local people interpreted by translators. Hernández provided names in the Nahuatl indigenous language spoken by various peoples of central America (including the Aztecs), as well as Latin. He divided the plants he encountered into woody and herbaceous, further splitting those according to their uses: edible, medicinal, ornamental and economic. Three indigenous artists – named as Antón, Baltazar Elías and Pedro Vázquez – provided illustrations of the plants (as well as the animals and minerals) that Hernández described. Hernández gathered information for the book between 1570 and 1577, but although several authors included extracts from the work in their own books, it was not until 1651 that a (condensed) version of the original was published. At the time it was the largest compendium of naturalistic knowledge of the so-called 'New World'. It contained

RIGHT

Historia naturalis Brasiliae, Georg Marcgrave and Willem Piso, (1648)

This title page from this book, which was the first scientific work on the natural history of Brazil, shows a curious combination of indigenous people and a rather classical looking river god.

HISTORIA NATVRALIS
BRASILIAE,
Auſpicio et Beneficio
ILLUSTRISS. I. MAVRITII COM. NASSAU
ILLIUS PROVINCIÆ ET MARIS SUMMI PRÆFECTI ADORNATA:
In qua
Non tantum Plantæ et Animalia, ſed et Indigenarum morbi, ingenia et mores deſcribuntur et
Iconibus ſupra quingentas illuſtrantur.

LUGDUN. BATAVORUM,
Apud Franciſcum Hackium
et
AMSTELODAMI,
Apud Lud. Elzevirium. 1648.

descriptions of several plants that would subsequently become important global commodities, such as maize, chilli, tomatoes and cacao (from which chocolate derives).

Between 1630 and 1654, the Dutch West India Company (*Westindische Compagnie*, WIC), a similar set up to VOC focused on Brazil, the Caribbean and North America, occupied part of north-east Brazil. The territory was centred around Pernambuco, which at the time was the greatest sugar producer in the world. Given jurisdiction over Dutch participation in the Atlantic slave trade, the WIC sought to profit from enslaving people, and exploiting sugar cane and Brazilian timber. Count Johan Maurits of Nassau-Siegen (1604–79), who governed Dutch Brazil between 1637 and 1644, gathered together a group of scholars and painters to explore the local geography, flora, fauna, indigenous people, tropical diseases and traditional medicine. Among them, the German naturalist Georg Marcgrave (1610–44) studied the plants, animals, geography, meteorology and astronomy of north-east Brazil, while the Dutch physician Willem Piso (1611–78) examined the local diseases and medicinal plants.

In 1648, WIC board member Johannes de Laet (1581–1649) transcribed, edited and published Marcgrave and Piso's work in Latin in *Historia naturalis Brasiliae* (*Natural History of Brazil*), depicting hundreds of plants with their vernacular names (using the language conventionally known as 'Tupi'), and outlining local ailments and treatments, and other uses of plants. Marcgrave observed that the fruit of the calabash tree was often used as a plate, cup or bottle, and that the indigenous people could make fire without stones using a piece of the stem or root of the ambaiba tree, for example. The work combined such observations with references to classical scholars Theophrastus and Dioscorides and the authors' contemporaries, including Francisco Hernández and Carolus Clusius. A study published in 2019 found that of the 378 species found in *Historia naturalis Brasiliae*, 256 (68 per cent) were useful, with similarities in how they were used for food and healing in the 17th century and in modern Brazil.

Elsewhere, the VOC was gaining ground. By the mid-1600s it had 50,000 employees and 10,000 soldiers, and was operating trading posts from the Persian Gulf to Japan. Travelling via the Cape of Good Hope and the Strait of Magellan, its 150 merchant ships made regular journeys from Holland to West and South Africa, Mauritius, Arabia, Persia, southern India, Ceylon (modern-day Sri Lanka), Java, the Moluccas (islands of north-east Indonesia), China and Japan. The company's main money-spinners were pepper, nutmeg and mace, cloves and cinnamon, but other plants were also of interest. Particularly sought-after were medicinal plants from Asia that could treat the tropical diseases that VOC staff frequently succumbed to. Those employed by VOC as gardeners, apothecaries and surgeons were instructed to collect information about any interesting plants they encountered in and around their stations. According to Herman Boerhaave (1668–1738) of the University of Leiden, 'practically no captain whether of a merchant ship or of a man-of-war, left our harbours without special instructions to collect everywhere seeds, roots, cuttings and shrubs and bring them back to Holland.'

The German VOC botanist Georg Eberhard Rumphius (1627–1702) studied the flora and fauna of Ambon (one of the Moluccan islands, where coveted spices grew) but faced considerable hurdles in his efforts to publish his *Herbarium amboinense* (*Herbarium of Ambon*). Having started work in 1662, he went blind in 1670 but continued to write

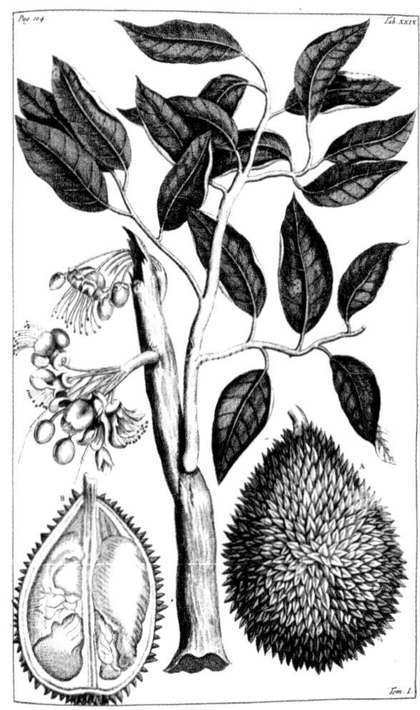

ABOVE

Herbarium amboinense, **Georg Eberhard Rumphius**, (1741–1750)

Rumphius described plants, such as this durian, growing on the island of Ambon in the Moluccas.

the manuscript with the help of his wife, Susanna. However, four years later, she and one of their daughters were killed during an earthquake and tsunami. The work continued with the help of several assistants and Rumphius' son, Paul August, who contributed many illustrations. Progress was thwarted once more in 1687 when a fire at Rumphius' library destroyed important documents. Then, after the work was finally completed in 1690 the ship carrying it to the Dutch Republic (modern-day Netherlands) was attacked by the French and sank, forcing a rewrite from a copy. When the manuscript finally arrived in the Dutch Republic in 1696, the VOC decided its contents were so commercially sensitive it should not be published. Altogether it profiled 1,300 flowering plant species, including cloves, which were used in Europe as a culinary spice and for treating ailments ranging from dizzy spells, to constipation and the plague. *Herbarium amboinense* was eventually published in Latin by Johannes Burman (1707–70), Professor of Botany and Medicine and Director of Amsterdam Botanic Garden between 1741 and 1755, half a century after Rumphius had died.

Johannes Burman was also instrumental in printing botanical information gathered by other VOC servants. In 1737, he published *Thesaurus Zeylanica* (*Treasures of Ceylon*), primarily founded on the work of VOC physician Paul Hermann (1646–95). Hermann had collected plants in Ceylon between 1672 and 1679, where 'of all the plants, which the learned botanist had seen before in Europe, he found in Ceylon hardly three specimens'. After Hermann died in 1695, his English botanist friend and pupil William Sherard (1659–1728) published Hermann's work 'in its original imperfect and chaotic state' as *Musaeum Zeylanicum* (*Museum of Ceylon*) in 1717.

The work of editing and reclassifying Hermann's work was left to Burman, aided by the Swedish naturalist Carl Linnaeus (1707–78) – who would go on to revolutionize plant classification and naming. The work describes more than 650 plants, including the Malabar nut, amaranth, cinnamon and different types of jasmine, with names given in Latin, Malabar (Tamil) and Singhalese. Burman similarly drew on collections made at the Cape of Good Hope by Hermann when en route to Ceylon, as well as those of VOC Cape garden superintendents Heinrich Bernhard Oldenland, Jan Hartog and Johann Andrea Auge, when he published *Rariorum Africanarum plantarum* (*Rare African Plants*) during 1738 to 1739. Containing one hundred descriptions of plants from the Cape of Good Hope, based on illustrations made by the physician Hendrik Claudius, it is one of the earliest works devoted to South African plants.

In India, Hendrik van Rheede (1636–91), the VOC's Governor of Dutch Malabar at Cochin from 1669 to 1676, was interested in plants and keen to study and codify the traditional medicines of southern India. In particular, he wanted to find alternative sources of medicine for Dutch troops, as drugs imported from Europe lost their potency on the long sea journey to India. In 1678, he initiated the first ever survey of tropical botany in South Asia. To this end, he mobilized a team of physicians, scientists, artists, plant collectors and interpreters, from both Malabar and Europe. Over many years, they produced the twelve-volume *Hortus Malabaricus* (*Garden of Malabar*), describing in Latin some 740 wild and cultivated plants, trees and shrubs of the Malabar coast, and outlining their medicinal properties. The plant names were given in Latin, Sanskrit, Malayalam and Konkani languages, with the Malayalam name also given in Arabic script. The text was illustrated with nearly 800 engravings, which were made in Holland based on drawings made in India by Dutch artists. The book was published in Amsterdam between 1678 and 1693, with the backing of the Raja of Cochin.

Local people were major providers of information for the book. Healers from the low-caste toddy-tappers community (toddy being wine made from the sap of particular palms) collected plants and provided information on how plants were used according to the Buddhist Ezhava tradition. The key source in this regard was the local Keralan herbalist Itty Achudan. Meanwhile, three Konkani Brahmin priest-physicians, Ranga Bhatt, Apu Bhatt and Vinayak Pandit, provided scholarly references and helped to write the text. Van Rheede recorded how orally transmitted information was agreed for inclusion in the report, recalling: 'I often attended a most delightful entertainment for instance when these Brahman and gentile philosophers disagreed and disputed with each

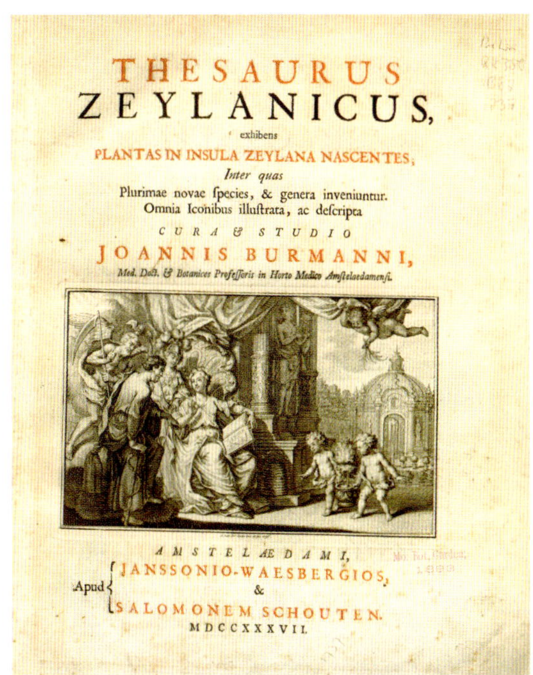

THESAURUS
ZEYLANICUS,
exhibens
PLANTAS IN INSULA ZEYLANA NASCENTES,
Inter quas
Plurimae novae species, & genera inveniuntur.
Omnia Iconibus illustrata, ac descripta
CURA & STUDIO
JOANNIS BURMANNI,
Med. Doct. & Botanices Professoris in Horto Medico Amstelaedamensi.

AMSTELÆDAMI,
Apud JANSSONIO-WAESBERGIOS, & SALOMONEM SCHOUTEN.
MDCCXXXVII.

BELOW

***Thesaurus zeylanicus*, Johannes Burman, (1737)**

This work was based on the collections of Paul Hermann, a medical officer for the Dutch East India Company (VOC).

RIGHT AND OVERLEAF

***Hortus Malabaricus*,**
Hendrik van Rheede,
(1678–1693)

The work included skilfully executed images detailing the flora on the Malabar Coast of southwestern India (this and next spread).

Fig. 5.
Caunga. lat.
Mal.
Arab.
bram.

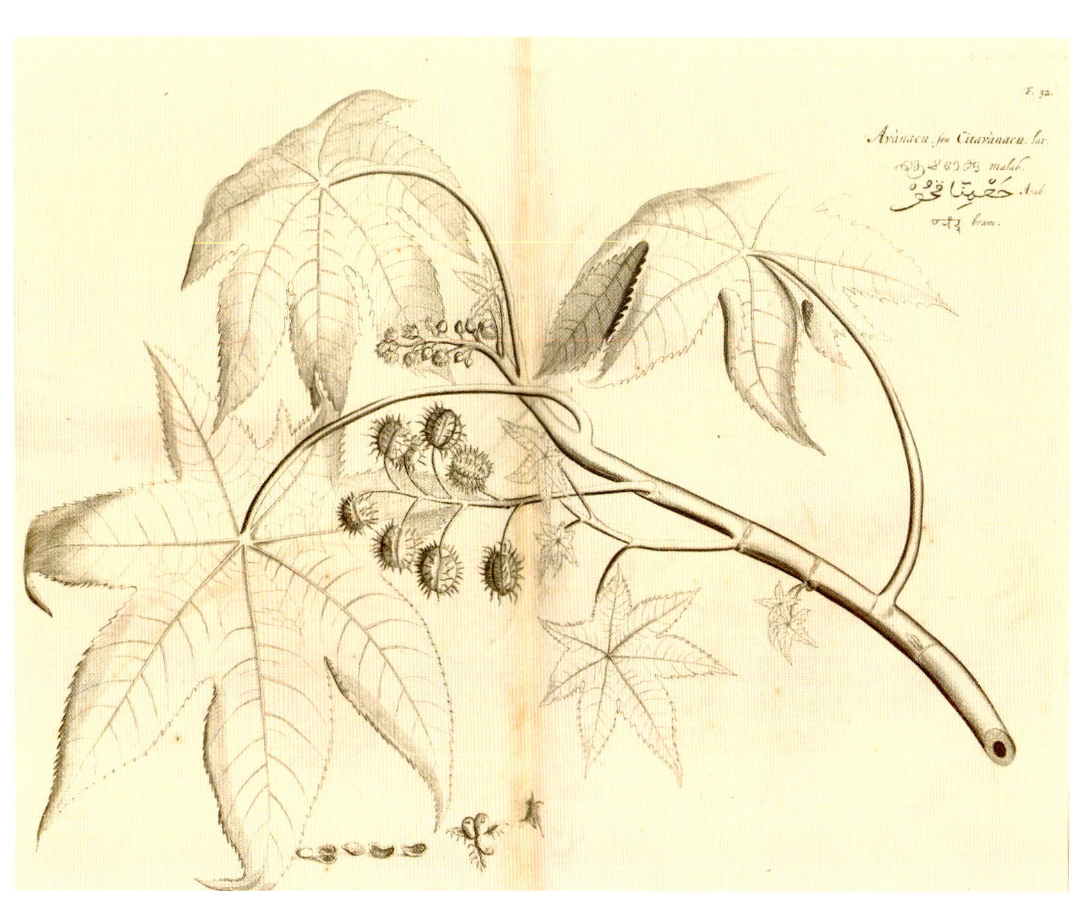
F. 32.
Avànacu, ſeu Citavànacu. lat.
malab.
Arab.
bram.

Fig 16
lat.
mal.
Arab.
bram.

other by weight of arguments, which they took from maxims, rules, verses from antiquity, and books of their ancestors who were renowned for their learning.'

The book *Icones Plantarum Malabaricum*, roughly contemporary to *Hortus Malabaricus*, contains 262 watercolour illustrations of medicinal plants from Sri Lanka, detailing local names, medicinal properties and uses. Most probably written by a VOC surgeon in the late seventeenth or early eighteenth century, during the Dutch occupation of Ceylon, it is another important ethnobotanical reference work for medicinal plants in South Asia.

4 Early depictions of Earth's web of biodiversity

Some of the botanical works produced by Europeans in the seventeenth century are notable for their examination of the interactions between animals and plants, hinting at an early awakening to wider ecological thinking. Among them are Maria Sibylla Merian's *Metamorphosis insectorum Surinamensium* (*Metamorphosis of Surinamese Insects*), published in 1705, which is one of the earliest works to feature the life cycles of insects and their food plants. Merian (1647–1717), a German naturalist, had learned copperplate etching by the age of eleven from her stepfather Jacob Marrel, who was a celebrated botanical illustrator. Within two years, she was combining her significant artistic skills with her interest in the natural world to document her efforts to raise silkworms on mulberry leaves. She became a flower painter, engraver, embroidery-pattern designer and teacher, and had already published three books showcasing flowers and two of butterflies when, in 1699, she decided to travel to Suriname, on the northeast coast of South America, accompanied by one of her two daughters.

After two months at sea they landed in Paramaribo, now the capital of present-day Suriname, which was then in the hands of the Dutch, who were profiting from the sugar trade using enslaved workers. Merian drew heavily on the knowledge and assistance of enslaved people working on Dutch plantations, which grew sugar cane, coffee and cotton, as she researched *Metamorphosis*. The finished book, which contained sixty engraved plates, showed the life cycles of butterflies, moths and other insects alongside plants and other animals, and included notes on diets and habitats. It featured several plants that were little known in Europe at the time, such as pineapples, prickly custard apples and guavas. Merian thought the pineapple was 'the most outstanding of all the edible fruit' and placed it on the first plate in the book. She included cockroaches on the image, noting that: 'Cockroaches are the most infamous of all insects in America on account of the great damage they caused to all inhabitants by spoiling their wool, linen, food and drinks.'

Following in Merian's footsteps in depicting plants and animals together was English naturalist Mark Catesby (1683–1749). Between 1729 and 1747, he published the first major illustrated account of British-occupied territories in southeastern North America. *The Natural History of Carolina, Florida and the Bahama Islands* comprised 220 plates and accompanying descriptions, of 'birds, beasts, fishes, serpents, insects, and plants: particularly, the forest-trees, shrubs, and other plants, not hitherto described, or very incorrectly figured by authors'. Catesby had first visited North America on a trip to Virginia between 1712 and 1719. He had amassed such a wealth of natural history

LEFT

Metamorphosis Insectorum Surinamensium,
Maria Sibylla Merian,
(1705)

Merian was pioneering in depicting insect life cycles and food plants together, as with the pineapple and cockroaches seen here.

RIGHT

The natural history of Carolina, Florida and the Bahama Islands, **Mark Catesby,** (1731–1743)

Catesby's 'handsomly cloathed' ... 'Bastard Baltimore Bird' and the Catalpa Tree.

OPPOSITE

The natural history of Carolina, Florida and the Bahama Islands, **Mark Catesby,** (1731–1743)

Catesby made this etching of a 'Blackcap flycatcher perched on yellow jessamy' in 1731.

T. 53
Jasminum luteum.
The yellow Jeſſamy.
Muscicapa nigrescens.
The blackcap Flycatcher.

specimens and drawings during the visit that a powerful group of London naturalists were persuaded to sponsor a second trip, this time to Carolina and Florida. Catesby's supporters included the English Royal Physician Hans Sloane, the diplomat and botanist William Sherard (who had worked on Hermann's *Musaeum Zeylanicum*) and Charles du Bois, Treasurer of the East India Company. In exchange for their patronage, Catesby sent them specimens for their collections and gardens.

When Catesby returned from this second visit, he set about writing up and illustrating his travels in the book that would become his life's work. Despite not being trained as a painter, he was artistically talented and produced precise visual records of specimens using 'a Flat tho' exact manner' that he considered 'may serve the Purpose of Natural History, better in some Measure, than in a more bold and Painter-like Way.' However, after meeting the botanical illustrator Georg Dionysius Ehret (1708–70), he adopted a more rounded and three-dimensional approach to highlight the details of seeds, buds, flower heads and roots. Like Merian, Catesby often presented flora and fauna together. For example, he illustrated the American goldfinch perched in an acacia tree, noting that such goldfinches were commonly kept in cages in New York. And he depicted the ground dove with a 'Pellitory or Tooth-ach Tree', explaining that the dove was about the size of a lark and ate the tree's berries, and that the people of coastal Virginia and Carolina used the aromatic seeds and bark of the tree to treat toothache, hence its name. Cromwell Mortimer, Secretary of London's Royal Society for Improving Natural Science (today known as the Royal Society), which had been founded in 1660, described the work in 1747 as the 'most magnificent work I know since the Art of printing has been discovered.'

ABOVE

***The natural history of Carolina, Florida and the Bahama Islands*, Mark Catesby, (1731–1743)**

This watercolour depicts an eastern king bird (called the tyrant bird by Catesby) on a branch of sassafras.

5 China and Japan embrace updated *materia medica*

While the indigenous people in South and Southeast Asia and the Americas were undergoing turbulent times at the hands of exploitative Europeans, the Hans, rulers of China's Ming dynasty since 1368, finally gave way to the north-east Asian Manchu people in the middle of the seventeenth century. During the Qing dynasty that ensued, China's territory and population grew, it established new crops (such as the potato, peanut and tomato) from the Americas, and became wealthy through export trade in tea and luxury products including silk and porcelain. Emperor Kangxi, who ruled between 1662 to 1722, oversaw cultural strides that included creating extensive maps of the country, and the continued production and circulation of knowledge. The extensive *Ben Cao Gang Mu* (*Compendium of Materia Medica*), which had been published late in the sixteenth century by the physician Li Shizhen (1518–93), played an important role in advancing understanding of medicinal plants across China and beyond. Recording 1,892 medicinal substances, and correcting errors from earlier *materia medica*, Li Shizhen had dropped the established toxicity grading and presented a new hierarchy of organisms. *Ben Cao Gang Mu* also circulated in Japan, where the founder of the Tokugawa regime, Tokugawa Ieyasu, encouraged the study of Chinese herbalism to underpin a lifestyle based on the teachings of Confucius. As a result, scholars studied the text throughout the Tokugawa period (1603–1868), and it helped nurture a local tradition of natural history.

6 China and Europe exchange botanical knowledge

During the seventeenth and early eighteenth centuries in China, the Ming and Qing dynasties responded to pressure from European exploration by maintaining a policy of isolation and exerting strict control over foreign trade, allowing it only through the port of Canton. As a result of this, as well as geographic and language barriers, the exchange of botanical information between Europe and China was limited. However, some plants made the journey west to be grown in European gardens. For example, the Chinese orange, at the time called 'Poma Aurantia', featured in Besler's *Hortus Eystettensis* of 1613. Jesuit scholars were the main channel for the exchange of scientific information between China and Europe in the seventeenth and early eighteenth centuries, valued by the Chinese for bringing in scientific books and knowledge from Europe, and writing about their observations for European audiences.

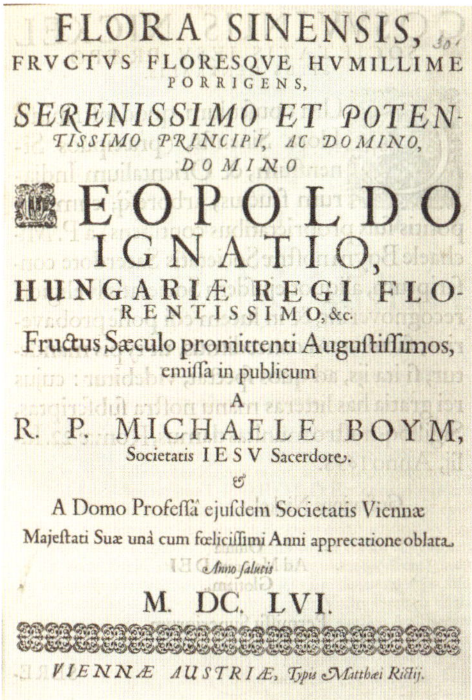

FLORA SINENSIS,
FRVCTVS FLORESQVE HVMILLIME PORRIGENS,
SERENISSIMO ET POTENTISSIMO PRINCIPI, AC DOMINO, DOMINO
LEOPOLDO IGNATIO,
HUNGARIÆ REGI FLORENTISSIMO, &c.
Fructus Sæculo promittenti Auguſtiſſimos, emiſſa in publicum
A
R. P. MICHAELE BOYM,
Societatis IESV Sacerdote.
&
A Domo Profeſſâ ejuſdem Societatis Viennæ
Majeſtati Suæ unà cum fœliciſſimi Anni apprecatione oblata.
Anno ſalutis
M. DC. LVI.
VIENNÆ AUSTRIÆ, Typis Matthæi Rictij.

BELOW

Michał Piotr Boym, (1612–59)

Boym's work *Flora sinensis* was one of the first books on China to be written by a European.

Flora sinensis (Flora of China), written by Polish Jesuit missionary and diplomat Michał Piotr Boym (1612–59) was highly influential in opening European eyes to plants growing in China. Published in 1656, it presented a selection of what Boym considered to be the 'principal fruits, trees and animals of the Chinese and Indians'. This encompassed both indigenous plants, including the lychee, and some that had been introduced from the Americas, such as the papaya, wild cashew and guava. The volume contains forty-eight pages of text and twenty-three black-and-white woodcut images (although some hand-coloured versions exist), of which seventeen are botanical, five are of animals and one is of a stele. Boym commented that not all plants were illustrated as some are 'in most of the herbariums'. For each plant, Boym gave a description, its names in Chinese and Latin, and information on where it grew; for some he also included medicinal properties. Interestingly, among the animals illustrated is the hippopotamus, which Boym could not have seen in China but he might have encountered on his inward journey, having travelled via Mozambique.

7 Japan and Europe follow parallel paths of learning

In the seventeenth and early eighteenth centuries, western Europe and Japan had much in common when it came to botany. Both cultures had been informed by classical systems of study, with Europe's understanding of plants rooted in Greco-Roman teachings and Japan's drawn from the classical Chinese tradition of *Ben Cao*. Independently, naturalists from both regions who were interested in botany began to shrug off their dependency on the past and undertake more autonomous and localized enquiries into the natural world. In both cultures, compilers of botanical knowledge began to collect accurate and detailed descriptions of plants, gather information from those who made practical use of plants and use accessible languages. Despite Japan, like China, having a strict policy of isolationism, the VOC was granted special trading privileges by the Tokugawa shogunate, and established a trading post in Nagasaki harbour. Botanical knowledge exchanged hands here, with Europeans gaining insights into a little-known country and Japan adapting European knowledge to suit its own needs.

BELOW

***Amoenitatum exoticarum* (*Exotic attractions*), Engelbert Kaempfer, (1651–1716)**

Kaempfer's work became the standard source of Western knowledge on Japan until the 19th century.

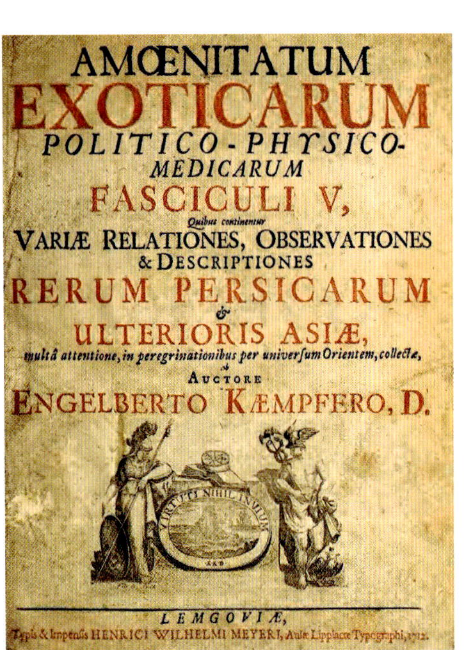
AMŒNITATUM
EXOTICARUM
POLITICO-PHYSICO-
MEDICARUM
FASCICULI V,
Quibus continentur
VARIÆ RELATIONES, OBSERVATIONES
& DESCRIPTIONES
RERUM PERSICARUM
&
ULTERIORIS ASIÆ,
multâ attentione, in peregrinationibus per universum Orientem, collectæ,
AUCTORE
ENGELBERTO KÆMPFERO, D.

LEMGOVIÆ,
Typis & Impensis HENRICI WILHELMI MEYERI, Aulæ Lippiacæ Typographi, 1712.

One particularly influential book that brought knowledge about Japan to Europe was by the German naturalist and physician Engelbert Kaempfer (1651–1716). Published in 1712, *Amoenitatum exoticarum* (*Exotic attractions*) included information on acupuncture, tea, amber, the nation's policy of seclusion and, importantly, an extensive *Flora Japonica* (*Flora of Japan*). Kaempfer had arrived at Nagasaki in 1690, equipped with a rare century-old copy of the Jesuit work *Thesaurus Linguae Japonicae* (*Japanese Thesaurus*), and stayed for two years. Despite the controls on Europeans in Japan, Kaempfer was able to make forays into Nagasaki and the surrounding mountains to collect herbs, and he twice visited the court at Edo (modern-day Tokyo), where he was warmly received. He considered the Japanese to be 'reasonable

LEFT

Flora sinensis,
Michał Piotr Boym,
(1656)

Boym described fruits and trees he encountered in China, including the lychee, seen here.

people and exceptional specialists and lovers of plants'.

In the *Flora Japonica* section, he presented 570 types of plants, 250 of which were illustrated with engravings. He grouped the plants as berry- and plum-bearing; apple-bearing and nut trees; oil-producing, fruiting and pod-bearing plants; plants with conspicuous flowers; and miscellaneous plants. Among those to feature were 'Laurus camphorifera', described as being related to laurel and the source of camphor oil; Kaempfer gave its Japanese name in characters and phonetic spelling, described the plant, and explained how camphor oil was extracted and used. Another useful plant was the varnish tree 'fafi no ki', the sap from which was used to make lacquer that the Japanese had long used to cover wood and metal items. Kaempfer is credited with introducing several plants to Europe, including ginkgo, which he observed under cultivation in a Buddhist monastery, and soya, which he noted was vital to Japanese cuisine.

8 Plants start to reveal their inner secrets

While many naturalists were engaged with exploring and gathering botanical intelligence from regions far from their home nations, other scientists were taking a more inward look at plants through the lens of a new technology: the microscope. It is not clear who was responsible for this invention but Dutch spectacle maker Zacharias Janssen (1585–*c.* 1632) – possibly aided by his father, Hans – is credited with making one of the first compound-lens microscopes around 1600. Some three decades later, an image of a plant viewed through a microscope was published for the first time in *De florum cultura* (*On the Cultivation of Flowers*), a work by Italian Jesuit gardener Giovanni Battista Ferrari (1584–1655). The book depicted the Rome garden of the Pope's nephew Cardinal Barberini, which Ferrari tended, and the image showed three views of a hibiscus seed.

A more comprehensive work that revealed the potential of microscopes to inform wider scientific investigation was *Micrographia*, published in 1665 by the English academic Robert Hooke (1635–1703). Hooke was the Curator of Experiments at the Royal Society in London and a gifted instrument-maker and scientist. In *Micrographia* he shared with the public the entirely novel world he encountered when peering through his compound microscope: among the entries were the detailed body segments of a flea; gravel in urine; the compound eye of a drone fly; mould growing on a leather-bound book; and fossilized wood.

Of particular significance to botanists were the microscopic depictions of a stinging nettle leaf and a sliver of cork. The magnified image in *Micrographia* of the former showed readers that the plant's sting resulted from a malign fluid that was drawn up through tiny needles on the leaf and into the skin. The image of the latter revealed cork's internal structure to be, 'all perforated and porous, much like a Honey-comb . . . these pores, or cells . . . were indeed the first microscopical pores I ever saw.' Hooke surmised these to be water-tight and air-filled, bestowing cork with its ability to float and be compressed to make bottle stoppers. Although the chambers he observed were not, in fact, biological cells but the walls remaining after living cells had died, it was the first use of the word 'cell' in this context. The English plant physiologist Nehemiah Grew later examined plants' morphology using a microscope, and, in his 1682 work *The Anatomy of*

LEFT

Robert Hooke,
(before 1633)

Painted by Mary Beale, this image is believed to be of Robert Hooke, author of *Micrographia.*

BELOW

Micrographia,
Robert Hooke,
(1665)

Micrographia presented images of cork, snowflakes, fleas, leaves and more, as viewed under the microscope.

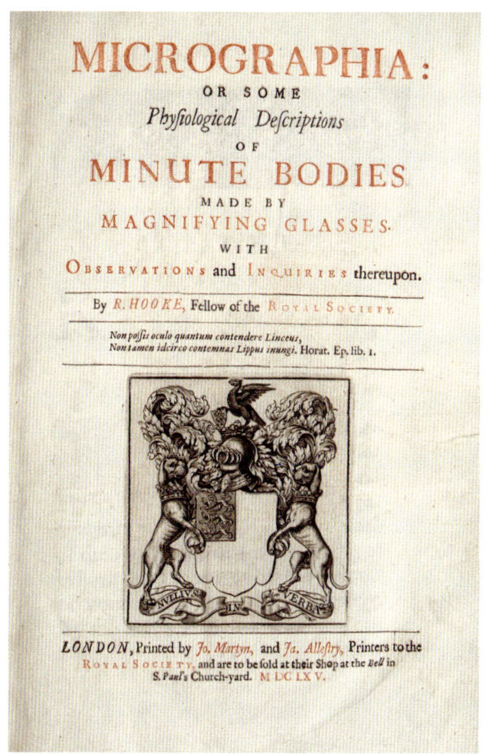

MICROGRAPHIA:
OR SOME
Physiological Descriptions
OF
MINUTE BODIES
MADE BY
MAGNIFYING GLASSES.
WITH
OBSERVATIONS and INQUIRIES thereupon.

By *R. HOOKE*, Fellow of the ROYAL SOCIETY.

Non possis oculo quantum contendere Linceus,
Non tamen idcirco contemnas Lippus inungi. Horat. Ep. lib. 1.

NVLLIVS IN VERBA

LONDON, Printed by *Jo. Martyn*, and *Ja. Allestry*, Printers to the ROYAL SOCIETY, and are to be sold at their Shop at the *Bell* in S. *Paul's* Church-yard. M DC LX V.

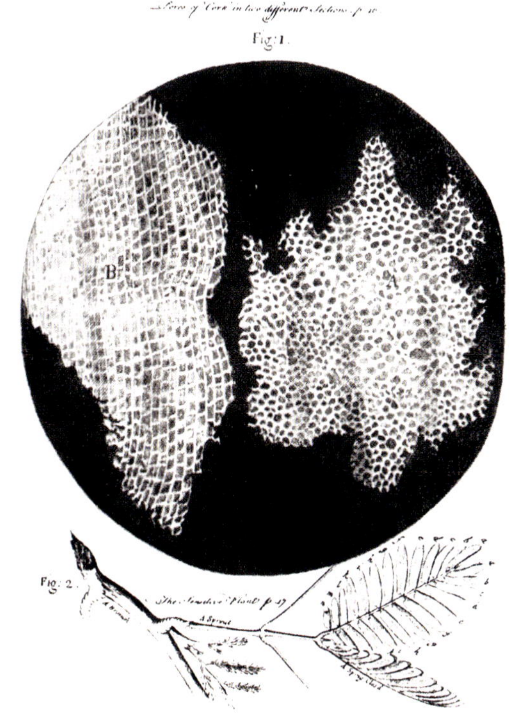

THE
ANATOMY
OF
PLANTS.
WITH AN
IDEA
OF A
Philosophical History of Plants,
And several other
LECTURES,
Read before the
ROYAL SOCIETY.

By NEHEMIAH GREW M.D. Fellow of the ROYAL SOCIETY, and of the COLLEGE of PHYSICIANS.

Printed by W. Rawlins, for the Author, 1682.

TOP

Nehemiah Grew,
(1641–1712)

Plant physiologist Nehemiah Grew pioneered the study of plant anatomy.

ABOVE

The Anatomy of Plants,
Nehemiah Grew,
(1682)

The 82 illustrated plates in his work revealed the inner structure of plants in meticulous detail.

ABOVE

Carl Linnaeus,
(1707–1778)

Linnaeus framed principles for defining genera and species, and created a system for naming them.

Plants, presented the first work on plant cell structure. The cell would later come to be recognized as the basic building block of all living organisms, including plants.

9 A way to classify and name all of nature

As the seventeenth century drew to an end, plants from around the world were thriving in European gardens, and preserved botanic material was beginning to amass in herbaria across the continent. However, many of the specimens were difficult to identify and name, as their characteristics simply did not fit into classification systems developed thus far. An important advance in understanding came in 1694, when Rudolf Jacob Camerarius (1665–1721), director of the botanical garden in Tübingen, Germany, observed that flowers with pistils would not set seed without the presence of flowers with stamens. In a letter to a professor at Giessen, he called the pistil a flower's female sex organs and the stamens its male organs, adding that pollen was needed for seeds to develop. This established for the first time that sexual reproduction took place in plants, and set the scene for the Swede Carl Linnaeus (1707–78) to develop an entirely new classification system based on this understanding.

Linnaeus, the son of a clergyman, had become interested in garden flowers as a child, diligently learning the names of the plants. After enrolling at the University of Uppsala, intent on learning all he could about the natural world, he was taken under the wing of the elderly professor Olof Rudbeck (1660–1740), who guided him to publish his first paper on the sexuality of plants in 1729. Three years later, Linnaeus set out to explore the botany of Lapland, travelling 7,600km (4,725miles) and gathering more than 500 specimens. This would be published in 1737 as *Flora Lapponica* (*Flora of Lapland*). In the meantime, Linnaeus had other priorities. In 1735, he headed to the Netherlands to complete his medical degree at Harderwijk. It was during his time in the country that he not only met Johannes Burman, for whom he worked on *Flora Zeylanica*, but also the physician Hermann Boerhaave, by now retired from the University of Leiden, and the Dutch botanist Jan Frederik Gronovius (1690–1762), who had a large herbarium and became Linnaeus' patron.

In 1735, Linnaeus published *Systema naturae* (*The System of Nature*), laying out a sexual classification for identifying plants, animals and minerals. In it, he divided each kingdom into classes, orders, genera, species and varieties. He determined the class to which a plant belonged according to the number and position of its stamens or 'husbands' (the male pollen-producing part of a flower), and placed it into an order based on the number and positions of its pistils, or 'wives' (the female reproductive part of a flower). Following Tournefort, he then split the orders into genera, and finally he divided each genus into species, differentiating each plant type from the others in its grouping. The language Linnaeus used outraged some of his contemporaries; while he suggested flower petals served as the 'marriage bed', the reality was that many plants reproduced in ways more akin to homosexuality, polygamy and incest than to heterosexual monogamy. One critic, the German physician and botanist Johann Siegesbeck, described Linnaeus' method of classifying plants as 'loathsome harlotry'. 'Who would have thought that bluebells, lilies and onions could be up to such immorality?' he mocked.

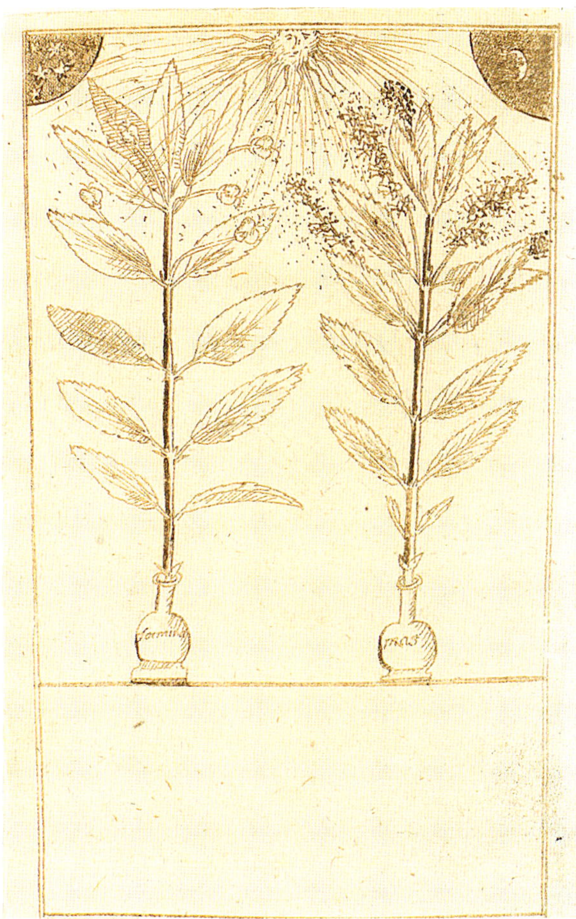

Caroli Linnæi
Medic: & Botan: Cult:
Stipend: Reg:
Præludia
Sponsaliarum
Plantarum
in quibus
Physiologia earum explicatur, Sexus demonstratur, modus generationis detegitur, nec non summa plantarum cum animalibus analogia concluditur
Upsal. 1729.

BELOW

Praeludia sponsaliorum plantarum,
Carl Linnaeus,
(1729)

Linnaeus' first paper was on the sexuality of plants, a subject that some considered blasphemous.

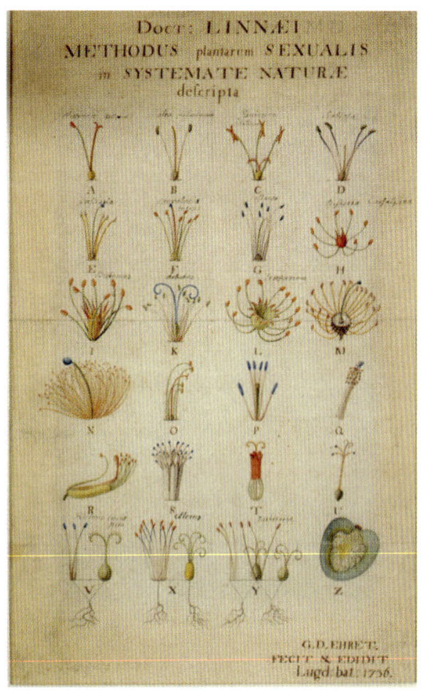

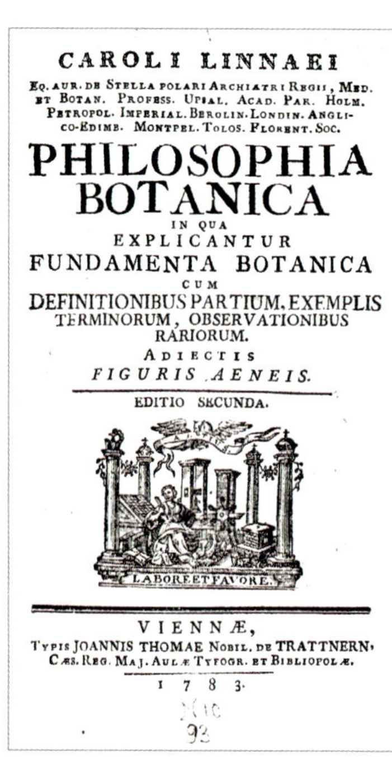

CAROLI LINNAEI

EQ. AUR. DE STELLA POLARI ARCHIATRI REGII, MED. ET BOTAN. PROFESS. UPSAL. ACAD. PAR. HOLM. PETROPOL. IMPERIAL. BEROLIN. LONDIN. ANGLICO-EDIMB. MONTPEL. TOLOS. FLORENT. SOC.

PHILOSOPHIA BOTANICA

IN QUA EXPLICANTUR FUNDAMENTA BOTANICA CUM DEFINITIONIBUS PARTIUM, EXEMPLIS TERMINORUM, OBSERVATIONIBUS RARIORUM.

ADIECTIS FIGURIS AENEIS.

EDITIO SECUNDA.

LABORE ET FAVORE.

VIENNÆ,

TYPIS JOANNIS THOMAE NOBIL. DE TRATTNERN, CÆS. REG. MAJ. AULÆ TYPOGR. ET BIBLIOPOLÆ.

1 7 8 3.

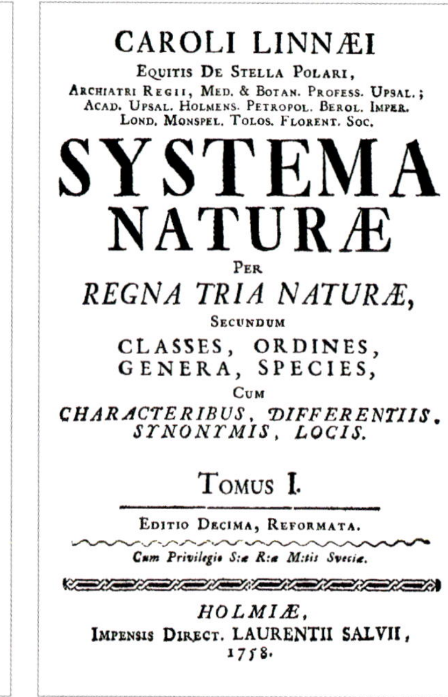

CAROLI LINNÆI

EQUITIS DE STELLA POLARI, ARCHIATRI REGII, MED. & BOTAN. PROFESS. UPSAL.; ACAD. UPSAL. HOLMENS. PETROPOL. BEROL. IMPER. LOND. MONSPEL. TOLOS. FLORENT. SOC.

SYSTEMA NATURÆ

PER REGNA TRIA NATURÆ,

SECUNDUM CLASSES, ORDINES, GENERA, SPECIES,

CUM CHARACTERIBUS, DIFFERENTIIS, SYNONYMIS, LOCIS.

TOMUS I.

EDITIO DECIMA, REFORMATA.

Cum Privilegio S:æ R:æ M:tis Sveciæ.

HOLMIÆ,

IMPENSIS DIRECT. LAURENTII SALVII, 1758.

ABOVE LEFT

Genera plantarum,
Carl Linnaeus,
(1737)

This drawing, by Georg Ehret, depicts Linnaeus' sexual system of plant classification.

ABOVE CENTRE

Philosophia botanica,
Carl Linnaeus,
(1751)

This work laid out Linnaeus' two-word system for naming species.

ABOVE RIGHT

Systema naturae,
Carl Linnaeus,
(1735)

This work explained Linnaeus' method for classifying three kingdoms - plants, animals and stones - into classes, orders, genera, species and varieties.

Through his association with Boerhaave, Linnaeus landed the job of house physician and gardener for the wealthy financier and VOC director George Clifford (1685–1770), which he undertook between 1735 and 1737. During this time, Linnaeus developed a series of principles and rules to be followed for classifying and naming plants. He published these in *Fundamenta botanica* (*The Foundations of Botany*) in 1736, and subsequently extended them in *Bibliotheca botanica* (*The Library of Botany*, 1736), *Critica botanica* (*A Critique of Botany*, 1737) and *Classes plantarum* (*Classes of Plants*, 1738). He then applied his theoretical framework in *Hortus Cliffortianus* (*Clifford's Garden*), a catalogue of the species in his employer's garden, published in 1737. In it, he took great delight in naming the hairy-leafed herb known as pig pungent weed as *Sigesbeckia orientalis*, after his critic Johann Siegesbeck. Linnaeus went on to publish *Genera plantarum* (*Genera of Plants*) the same year. This described the 935 plant genera known to Linnaeus at that time; he made several later revisions, as plants with new characteristics came to his notice.

Having returned to Sweden in 1738, Linnaeus became the chair of medicine and botany at his *alma mater* the University of Uppsala in 1742. As within other European countries, science and commerce were closely entwined at this time. So, it was natural that the Royal Swedish Academy of Sciences, which Linnaeus helped to found in 1739, and the decade-old Swedish East India Company (SEIC) would be allies. Over a prolonged period, Linnaeus sent seventeen of his best students abroad with the SEIC. The first was Christopher Tärnström (1711–1746), who travelled to China with instructions to bring back natural history specimens in general, and a living tea plant and seeds of mulberry trees in particular. Linnaeus believed that if his nation could grow commodities that fuelled economies elsewhere, Sweden itself could become wealthier. Two more men went to China at Linnaeus' request, while others visited South Africa, India, North America and South America. The students sent to South America included Daniel Solander, who would later participate with Joseph Banks in James Cook's first circumnavigation of the world. Wherever Linnaeus' students went, they spread the word of his revolutionary approach to classifying and naming plants.

ABOVE
Olof Rudbeck, (1660–1740)

Rudbeck, a professor of medicine at the University of Uppsala, Sweden, supported Linnaeus as a student.

In 1751, a year after becoming rector of the University of Uppsala, Linnaeus published *Philosophia botanica* (*The Philosophy of Botany*), which expanded on *Fundamenta botanica* and *Critica botanica* to present the first full statement of Linnaeus' binomial nomenclature. He did not invent the two-part Latin name, but he explained how the form could capture a plant's (or other organism's) unique position within its genus. Linnaeus' main goal was to standardize plant naming, so that correspondents far removed from each other could ensure they were discussing the same plant. 'If you do not know the names of things, the knowledge of them is lost too,' he warned. Prior to this work, a plant could sometimes have several names that were made up of the same words in different orders simply because botanists were divided over which was the more important characteristic of 'yellow flowered' or 'prickly leaved'. Now, with a foolproof way to classify plants and a simple way to name them, Linnaeus just needed to apply his technique to actually name all the plants he knew of.

Linnaeus scrutinized the specimens his botanizing 'apostles' brought back, and also consulted the books of the day. For example, he used Ray's work for classifying the British flora, Rheede's *Hortus Malabaricus* as the basis for the nomenclature he applied to Indian plants, and the *Flora Zeylanicum* files he had worked on for Burman when considering plants of Ceylon. He also consulted Rumphius' work on tropical botany, and Kaempfer's *Flora Japonica* for Japanese species. In 1753, he was finally able to publish *Species Plantarum* (*The Species of Plants*), listing all the species of plants known at that time and placing them in genera using the sexual system of *Systema naturae*. Naming 5,940 species, this first edition of the work, along with the 1754 revision of *Genera plantarum*, formed the basis for modern taxonomy, rendering earlier names as obsolete. Linnaeus, whose motto was '*Deus creavit, Linnaeus Disposuit*' (God created, Linnaeus organized), had fulfilled his life's ambition. It would be up to others to apply his system to classify and officially name the plants that botanists had yet to encounter, and to unravel the mechanism that gave rise to the diversity of plants across the world.

4

THE GLOBAL AND THE LOCAL
(1750–1830)

The Enlightenment from the mid-eighteenth century saw the urge for classification and organization of knowledge coming to predominate in the world of natural history. Naturalists now understood the world in terms of species, and wanted a practical, usable framework to organize what they saw. They also had Latin terms and definitions for each part of the plant, and taxonomic descriptions could be exchanged successfully. This required naturalists to accept there was a basic plant 'body type' and that even where parts looked strikingly different, they were nonetheless the same. It has allowed botanists to look, for example, at a cactus, and begin to see that its spines, produced from the stem, are adapted leaves.

1 Linnaean methods span the globe

Carl Linnaeus' (1707–78) ideas (see also page 134) spread rapidly for several reasons. One was that his previous research and publications gave him huge authority within the world of natural history. Another was that he had always worked tirelessly to create networks with like-minded scholars, spring-loading the value of using his systems. Finally, his naming methods worked, simplifying and streamlining the way that names were given and utilized. His classification, based on the sexual parts of the flowers, would become extremely popular because it was so usable.

Lady Anne Monson (1726–76) helped to translate Linnaeus' work into English for the first time. As an aristocratic young woman of keen intelligence, Monson was well-educated and had plenty of time to devote to this project. Monson's name did not appear on the translation, though. The author on the title page of *An Introduction to Botany: containing an explanation of the theory of that science, extracted from the works of Dr Linnaeus* (1776), was that of Hammersmith nurseryman and botanist James Lee (1715–95). Lee began work on Monson's suggestion that a version in English would be useful to those who had not learnt Latin. The book comprised an edited selection of writing from various of Linnaeus' works, possibly aimed at a female audience. Linnaeus' system based on sexual characteristics was considered likely to cause renewed offence when translated from Latin into the bald reality of English – Monson and Lee followed Linnaeus in describing male stamens as 'husbands' and female pistils as 'wives'.

BELOW

An Introduction to Botany,
James Lee, 1776

Title page of this work; Lee was greatly assisted by Lady Anne Monson, who helped to translate Linnaeus' ideas from Latin into readable English.

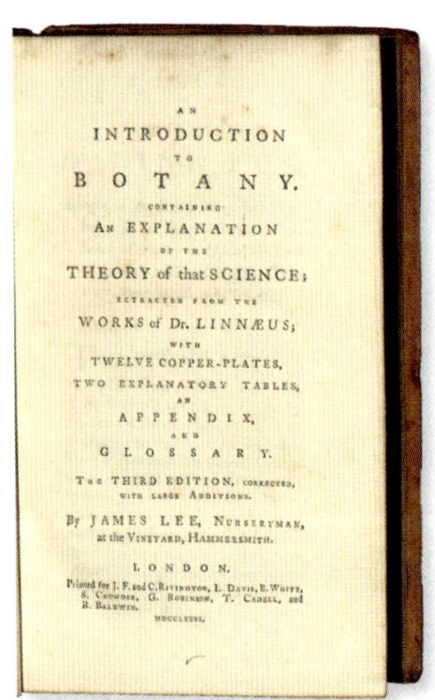

AN
INTRODUCTION
TO
BOTANY.
CONTAINING
AN EXPLANATION
OF THE
THEORY of that SCIENCE;
EXTRACTED FROM THE
WORKS of Dr. LINNÆUS;
WITH
TWELVE COPPER-PLATES,
TWO EXPLANATORY TABLES,
AN
APPENDIX,
AND
GLOSSARY.
THE THIRD EDITION, CORRECTED,
WITH LARGE ADDITIONS.
By JAMES LEE, NURSERYMAN,
at the VINEYARD, HAMMERSMITH.
LONDON,
Printed for J. F. and C. RIVINGTON, L. DAVIS, B. WHITE, S. CROWDER, G. ROBINSON, T. CADELL, and R. BALDWIN.
MDCCLXXVI.

James Lee's expertise in exotic plants, gained as a gardener at Syon House in Isleworth and then running his nursery, next ensured his recruitment to work on the first catalogue of plants at the Royal Botanic Gardens at Kew. Meanwhile his Linnaeus edition was so successful that it cornered the market in English until the end of the eighteenth century and beyond. In 1810, his son (also called James Lee) continued the franchise. The plates from James Lee junior's edition demonstrate in vivid, simple illustrations the distinctions Linnaeus was trying to draw between the various parts of the plant.

Lee formed part of Linnaeus' wide correspondence network, for example writing to Uppsala in 1776 to send a new species of mesembryanthemum brought from Madeira by the Kew plant collector Francis Masson (1741–1805). In the same letter Lee describes Joseph Banks and Daniel Solander's collections: 'certainly the greatest & I believe the best that ever was collected'. Banks' influence on botany can hardly be overstated.

2 Joseph Banks and his circle

Joseph Banks (1743–1820) was the son of a Lincolnshire landowner, and divided his time between there and central London. At the University of Cambridge the young Banks became impatient with the curriculum (intended to prepare churchmen) and paid privately for botanical instruction, exhibiting a proactive energy that would distinguish his whole life. He also began visiting the British Museum, where he met Linnaeus' student Daniel Solander (1733–82), recently moved to London to catalogue the museum's natural history collections. Solander would go on to invent the Solander Box, which is still valued in archives and libraries today as the best way of storing herbarium sheets and other loose materials.

Banks and Solander were recruited for the circumnavigation expedition, led by Captain James Cook during 1768–71, and on their return intended to publish their extensive collection of some 30,000 specimens in a florilegium. Their Southern Hemisphere collections comprised about 1,300 new plant species – an increase of a quarter on those known at the time. However, the travellers suffered some of the significant obstacles of the era: their expedition artist Sydney Parkinson died of dysentery before the voyage was over, which meant that Banks needed to find and pay new artists to complete his work, and then after that a team of eighteen engravers was enlisted to make 743 copperplates from these watercolour paintings.

The illustrations were almost complete when Solander died of a stroke in 1782, at which point Banks ceased work on the project altogether. This was a blow for Solander's posthumous reputation, almost certainly greater had the project been completed. The florilegium of the Australian trip was eventually published between 1980 and 1990 by the Natural History Museum, London, in a hundred sets at a price of about $100,000. The unfinished nature of these ventures reminds us that for all the treasured volumes contained in this Botanist's Library, there are many others that did not make it.

Another project sponsored by Banks was the journey of the Forsters, George (1754–94) and his father Johann (1729–98), German naturalists and artists who travelled on the second voyage of Captain James Cook, 1772–75. On their return, they produced *Characteres generum plantarum* (*The Characteristics of Plants*, 1775–76), the first description of the flora of New Zealand. The book's seventy-eight plates illustrate flowering plants at almost life-size, and the Forsters took the opportunity of having so many new species to name many after friends and supporters. It is noticeable that very many of the newly named species of the time recognize establishment figures resident in Europe, and rarely honour the local collectors and knowledge holders who had enabled their collection.

In later life, Banks did not do much new scientific work of his own but instead poured energy into the running of London's Royal Society, its oldest scientific institution, the creation of Kew Gardens, and the expansion of the British Museum. He paid and despatched many knowledge and specimen collectors across the globe; he also funded activities closer to home, such as William Smith's geological map of England, an extraordinary and unprecedented piece of work, and supported exploration by citizens of other nations such as the German Alexander von Humboldt (see also page 162).

ABOVE

Lady Anne Monson, unknown artist

Monson was born at Raby Castle, Co. Durham, and was educated in aristocratic fashion in the gardens and library; she was recognized as an expert botanist by her peers.

BELOW

***Botanic Manuscript*, Jane Colden, *c.* 1750s**

This work, now held in the British Museum, is a beautiful example of an amateur making a useful collection of local species, here, those of the Hudson Valley in New York.

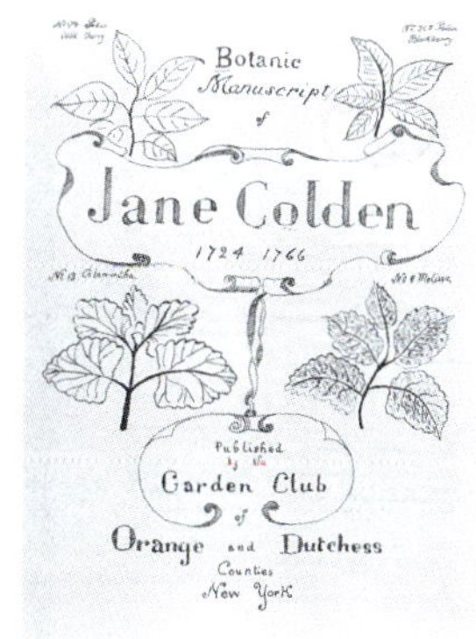

ABOVE LEFT

Self-portrait of Sydney Parkinson, date unknown

This simple painting does not hint at the sophisticated images of plants produced by Parkinson in difficult conditions while travelling with Joseph Banks.

ABOVE RIGHT

Joseph Banks, by Sir Joshua Reynolds, (*c.*1771–3)

Banks' name was listed in Reynolds' sitter book several times between 1771 and 1773; the portrait captures with great liveliness Banks' energetic, determined and open character.

However, he also used botanical knowledge in less edifying ways, hoping to transplant breadfruit to the Caribbean to feed enslaved people there more cheaply. He epitomizes the dilemma we have today in judging such a life.

3 William Curtis, the study of local flora, and the widening of botanical enthusiasm

At the same time as Banks and his colleagues were reaching out across the world, adding thousands of new species to the catalogues, another set of naturalists were following a different set of predecessors in paying attention to the very local. William Curtis (1746–99) produced his *Flora Londonensis* (*The Flora of London*) from 1777 in six large volumes, with plates by many distinguished botanical artists of the time. Curtis had begun his working life as an apothecary, writing his first book about collecting and preserving insects at the same time as becoming demonstrator and *Praefectus Horti* at the Chelsea Physic Garden, at that time on the country edge of western London. Curtis was hard-working and ambitious, clearly hoping he took the temperature of his time by writing his compendious guide to the flora growing within a 16km (10 mile) radius of the heart of the city. For the first time, such a book was aimed at the general reader, published in

BELOW LEFT

Barringtonia, *Icones animalium et plantarum,* J F Miller,
(1776)

Botanical artist John Frederick Miller's illustration of the Barringtonia. The plant was named after English lawyer and naturalist Daines Barrington (1727–1800).

BELOW RIGHT

African corn lily, J F Miller,
(1776)

Miller was one of the team of artists who worked for Joseph Banks producing fine hand-coloured engravings from sketches made by Sydney Parkinson, depicted left.

THIS PAGE

***Leptosporum collinum*,**
George Forster,
unknown date

This plant, native to Polynesia, now called Metrosideros, hints at its frothy flowers emerging from the top of the stem. The painting was made after Forster went on Cook's second voyage.

OPPOSITE

***Metrosideros collina*,**
Sydney Parkinson,
(1768–1771)

This finished Parkinson watercolour shows the detail and vivacity of his work. Both these species are now considered to be one, showing the difficulties involved in taxonomy, then as now.

Metrosideros spectabilis.
Sydney Parkinson pinx 1769.
H.V.4 f.93.

ABOVE

Johann Forster, painted by Anton Graff, (1775)

Johann, a naturalist, and his son Georg travelled with Banks and Cook and kept scrupulous records of what they saw, as well as making extensive collections.

LEFT

View of HMS *Resolution* and *Discovery* at anchor in Huaheine, John Cleveley, (*c.* 1780)

Cleveley was a London-based engraver and helped to work up Banks' expedition drawings; this South Sea image was composed from several original drawings for an enthusiastic market. Cleveley turned the drawings from Cook's second voyage into engravings.

OPPOSITE

***Hibiscus meraukensis*, Frederick Polydore Nodder, (1778)**

This delicate watercolour of this southern hemisphere plant was made from one of Parkinson's original drawings by Nodder, a London-based illustrator.

English, and each volume came with a small number of hand-coloured prints, such as exquisite *Scolopendrium vulgare* ferns, and glorious wild honeysuckle. Although explicitly London in orientation, with the city then much more surrounded by countryside than it is today, the book in practice identified many of the common flora of south-east England.

However, the work was not financially successful, and Curtis was saved from bankruptcy by starting *Curtis' Botanical Magazine* (1787–), which remains today the longest running gardening magazine. The periodical rode the wave of enthusiasm for magazines and reviews that exploded around the turn of the nineteenth century, facilitated in part by new much more economical methods of papermaking and then colour printing. The magazine was a hit with plant enthusiasts who could not afford to buy whole volumes of expensive hand-coloured copperplate illustrations, but who nonetheless could afford to spend on one or two. Bound volumes of *Curtis'* fetch healthy prices at auction, even today.

While Curtis' efforts focused on London in 1777, the Reverend John Lightfoot (1735–88) was working with another London publisher to prepare his guide to Scotland, *The Flora Scotica, or a systematic arrangement, by the Linnean Method, of the native plants of Scotland and the Hebrides*. Lightfoot was employed by Margaret Bentinck, Duchess of Portland (1715–85), as a librarian and chaplain, managing her private museum at Bulstrode Hall, the largest private natural history collection in England. His study of Scottish plants had to be achieved during short trips, but the book did include a list of

BELOW AND OPPOSITE PAGE

An Illustration of the Sexual System of Linnaeus,
John Miller,
(1779)

This smaller pocket-sized edition of Miller's work translating Linnaeus' system into English appeared in response to readers asking for a book for use 'visiting the Gardens, or ranging the Fields'.

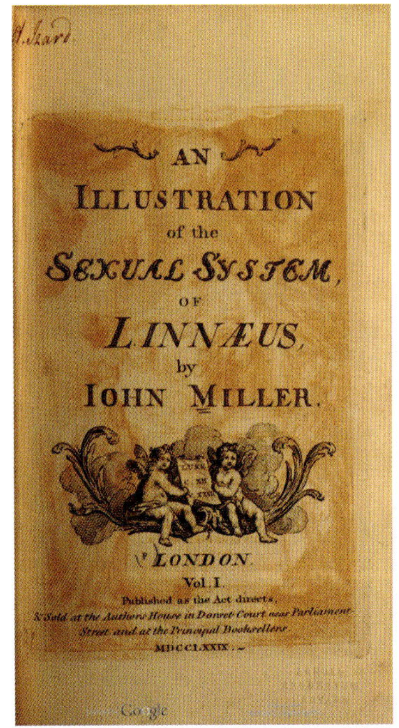
AN
ILLUSTRATION
of the
SEXUAL SYSTEM,
OF
LINNÆUS,
by
IOHN MILLER.
LONDON.
Vol. I.
Published as the Act directs,
& Sold at the Authors House in Dorset Court near Parliament Street and at the Principal Booksellers.
MDCCLXXIX.

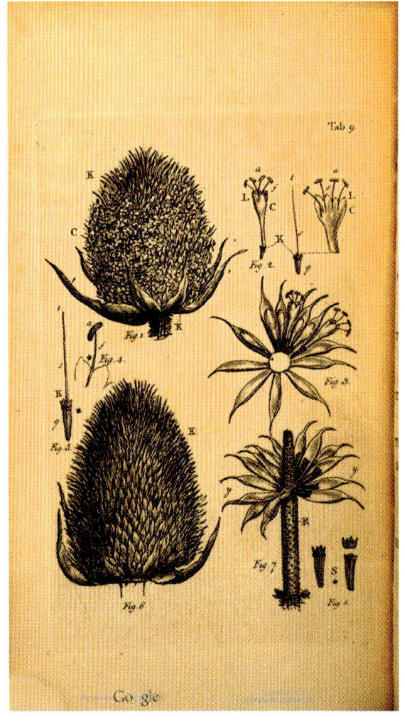

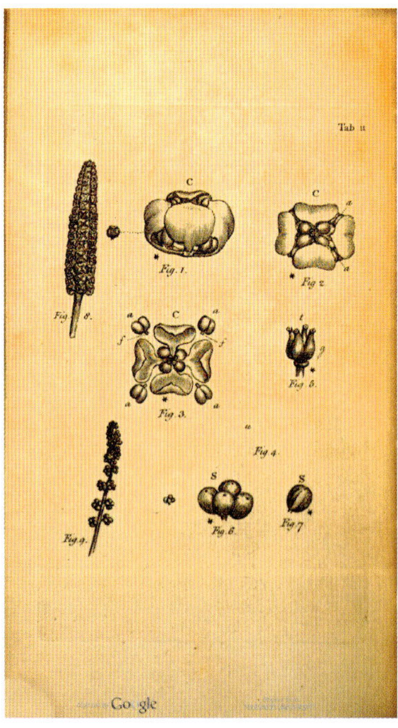

the Gaelic names for the plants alongside the English. (A similar volume he began writing for Wales was never published.)

An even more ambitious project was the first complete *English Botany* (1790–1814). This was the work of the botanist James Edward Smith (1759–1828), who founded the Linnean Society in London, acquiring for it his treasure house of papers on his death. However, Smith's name did not appear on the book, and it became known after James Sowerby (1757–1822) who had conceived the original idea and created all the illustrations. The work eventually comprised an astonishing thirty-six volumes with more than 2,500 plates.

In its original binding it is an extremely elegant addition to a bookshelf (perhaps it might have required two), but the illustrations show the way that botanical art was starting to focus on identifying details as well as depicting the whole plant, filling the white parts of the page with seedcases and other parts of the life cycle, and sometimes even anatomical cut-throughs. For example, Sowerby illustrates the structure that is hidden under the huge white petals of the waterlily, there to trap pollinators overnight; and the bindweed, *Convolvulus*, where the spiralling tendrils float away into the top corner of the picture.

The illustrations are only partly coloured, a technique that suggests the colour is there to be indicative rather than decorative, although it also substantially reduced the cost. James Edward Smith's name would eventually be added to his own title page, once he had

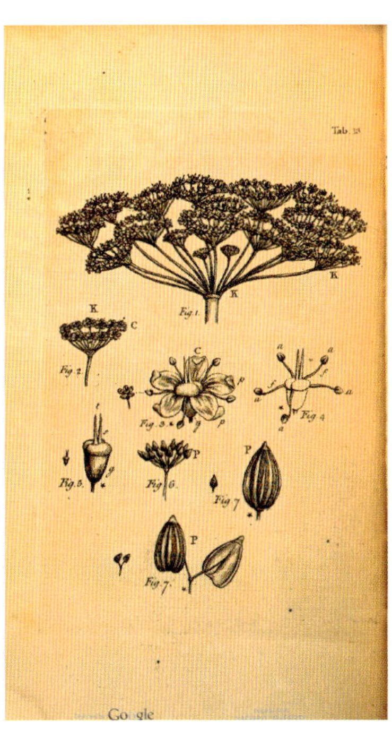

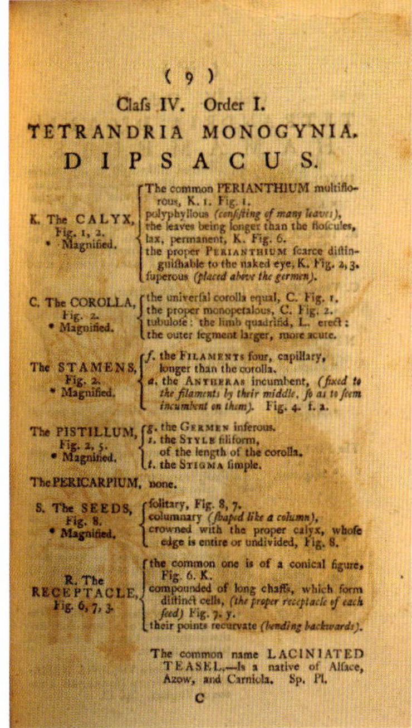

(9)

Claſs IV. Order I.

TETRANDRIA MONOGYNIA.

D I P S A C U S.

K. The CALYX, Fig. 1, 2. * Magnified.	The common PERIANTHIUM multifloROUS, K. 1. Fig. 1. polyphyllous *(conſiſting of many leaves)*, the leaves being longer than the floſcules, lax, permanent, K. Fig. 6. the proper PERIANTHIUM ſcarce diſtinguiſhable to the naked eye, K. Fig. 2, 3. ſuperous *(placed above the germen)*.
C. The COROLLA, Fig. 2. * Magnified.	the univerſal corolla equal, C. Fig. 1. the proper monopetalous, C. Fig. 2. tubuloſe: the limb quadrifid, L. erect: the outer ſegment larger, more acute.
The STAMENS, Fig. 2. * Magnified.	*f*. the FILAMENTS four, capillary, longer than the corolla. *a*. the ANTHERAS incumbent, *(fixed to the filaments by their middle, ſo as to ſeem incumbent on them)*. Fig. 4. f. a.
The PISTILLUM, Fig. 2, 5. * Magnified.	*g*. the GERMEN inferous. *s*. the STYLE filiform, of the length of the corolla. *t*. the STIGMA ſimple.
The PERICARPIUM,	none.
S. The SEEDS, Fig. 8. * Magnified.	ſolitary, Fig. 8, 7. columnary *(ſhaped like a column)*, crowned with the proper calyx, whoſe edge is entire or undivided, Fig. 8.
R. The RECEPTACLE, Fig. 6, 7, 3.	the common one is of a conical figure, Fig. 6. K. compounded of long chaffs, which form diſtinct cells, *(the proper receptacle of each ſeed)* Fig. 7. y. their points recurvate *(bending backwards)*.

The common name LACINIATED TEASEL.—Is a native of Alſace, Azow, and Carniola. Sp. Pl.

C

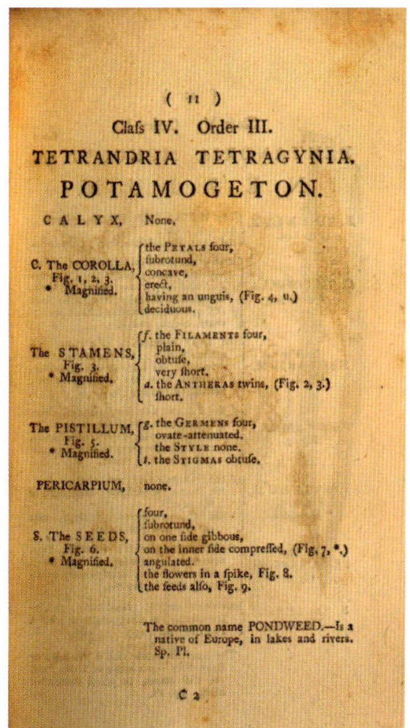

(11)

Claſs IV. Order III.

TETRANDRIA TETRAGYNIA.

POTAMOGETON.

CALYX,	None.
C. The COROLLA, Fig. 1, 2, 3. * Magnified.	the PETALS four, ſubrotund, concave, erect, having an unguis, (Fig. 4, u.) deciduous.
The STAMENS, Fig. 3. * Magnified.	*f*. the FILAMENTS four, plain, obtuſe, very ſhort. *a*. the ANTHERAS twins, (Fig. 2, 3.) ſhort.
The PISTILLUM, Fig. 5. * Magnified.	*g*. the GERMENS four, ovate-attenuated. the STYLE none. *t*. the STIGMAS obtuſe.
PERICARPIUM,	none.
S. The SEEDS, Fig. 6. * Magnified.	four, ſubrotund, on one ſide gibbous, on the inner ſide compreſſed, (Fig. 7, *.) angulated. the flowers in a ſpike, Fig. 8. the ſeeds alſo, Fig. 9.

The common name PONDWEED.—Is a native of Europe, in lakes and rivers. Sp. Pl.

C 2

RIGHT

Flora londinensis, or Plates and descriptions of such plants as grow wild in the environs of London,
William Curtis,
(1777)

Curtis' heavily Linnean flora effectively acted as a guide to the wild flowers of all southeastern England.

FLORA LONDINENSIS;

OR,

PLATES AND DESCRIPTIONS

OF SUCH

PLANTS

AS GROW WILD IN THE

ENVIRONS OF LONDON;

WITH THEIR

Places of *Growth* and Times of *Flowering*; their feveral *Names* according to LINNÆUS and other Authors:

WITH

A particular DESCRIPTION of each PLANT in LATIN and ENGLISH.

TO WHICH ARE ADDED,

Their feveral Ufes in *Medicine*, *Agriculture*, *Rural Œconomy*, and other *Arts*.

By WILLIAM CURTIS.

VOL. I.

LONDON:

Printed for and Sold by the AUTHOR, at his BOTANIC-GARDEN, *Lambeth-Marsh*; and B. WHITE and SON, Bookfellers, in *Fleet-Street*.

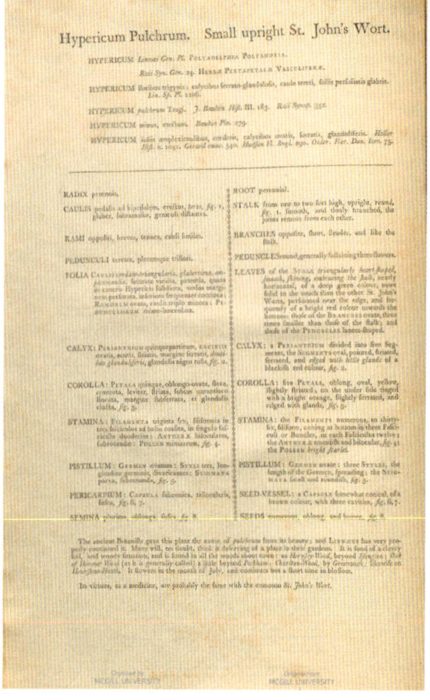

Hypericum Pulchrum. Small upright St. John's Wort.

HYPERICUM *Linnæi Gen. Pl.* POLYADELPHIA POLYANDRIA.

Raii Syn. Gen. 24. HERBÆ PENTAPETALÆ VASCULIFERÆ.

HYPERICUM floribus trigynis: calycibus ferrato-glandulofis, caule tereti, foliis perfoliatis glabris. *Lin. Sp. Pl.* 1106.

HYPERICUM *pulchrum* Tragi. *J. Bauhin Hift.* III. 183. *Raii Synop.* 342.

HYPERICUM minus, erectum. *Bauhin Pin.* 279.

HYPERICUM foliis amplexicaulibus, cordatis, calycibus ovatis, ferratis, glanduliferis. *Haller Hift.* n. 1041. *Gerard emac.* 540. *Hudfon Fl. Angl.* 290. *Oeder. Flor. Dan. Icon.* 75.

RADIX perennis.

ROOT perennial.

STALK from one to two feet high, *upright*, *round*, *fig.* 1. fmooth, and thinly branched, the joints remote from each other.

BRANCHES oppofite, fhort, flender, and like the ftalk.

PEDUNCULI teretes, plerumque triflori.

PEDUNCLES round, generally fuftaining three flowers.

LEAVES of the STALK *triangularly heart-fhaped*, *fmooth*, *fhining*, *embracing the ftalk*, nearly horizontal, of a deep green colour, more folid to the touch than the other St. John's Worts, perforated near the edge, and frequently of a bright red colour towards the bottom: thofe of the BRANCHES ovate, three times fmaller than thofe of the ftalk; and thofe of the PEDUNCLES lancet-fhaped.

CALYX: a PERIANTHIUM divided into five Segments, the SEGMENTS oval, pointed, ftriated, ferrated, and *edged with little glands* of a blackifh red colour, *fig.* 2.

COROLLA: five PETALS, oblong, oval, yellow, flightly ftriated; on the under fide tinged with a bright orange, flightly ferrated, and edged with glands, *fig.* 3.

STAMINA: the FILAMENTS numerous, to thirty-fix, filiform, uniting at bottom in three Fafciculi or Bundles, in each Fafciculus twelve; the ANTHERÆ roundifh and bilocular, *fig.* 4; the POLLEN *bright fcarlet*.

PISTILLUM: GERMEN ovate: three STYLES, the length of the Germen, fpreading: the STIGMATA fmall and roundifh, *fig.* 5.

SEED-VESSEL: a CAPSULE fomewhat conical, of a brown colour, with three cavities, *fig.* 6, 7.

In virtues, as a medicine, are probably the fame with the common *St. John's Wort*.

been knighted, admitted to the academies of Stockholm, Turin, Lisbon, Philadelphia, New York and other cities, and elected president of the Linnean Society in London: the combination of useful images and thorough Linnean description was a success. Certainly the preface suggested a moral justification for the book, as walking to botanize 'calls forth to healthy exercise the bodily as well as the mental powers',as scientific knowledge came in the late eighteenth and early nineteenth century to be seen as linked strongly with 'self-improvement'.

ABOVE

Groundsel, from *The Vegetable System*, Sir John Hill, (1759)

Produced under the patronage of Lord Bute, this huge botanical work with 1600 engravings was one of the first to use the newly devised Linnaean system of nomenclature.

4 The rise of French botany at the Jardin des Plantes

Similar efforts towards creating national floras were being made in other European countries. In France, Nicolas Francois Regnault (1746–*c.* 1810) and his wife Geneviève de Nangis Regnault (b. 1746) created over a twelve-year span from 1774 *La Botanique Mise à la Portée de Tout le Monde* (*Botany Made Available to Everyone*), which has been described by Wilfred Blunt as arguably the most impressive French botanical book of the period. The text was explicitly practical, linking it in some ways to the botany of the previous centuries, and it included a Table of Maladies to find specific cures for named sufferings. Geneviève has been particularly singled out for her skilled work in colouring the images, a job that often seems to have been reserved for women.

The British trend for native guides was also echoed over the Channel, in the *Herbier de la France* (*The French Herbarium*) by Pierre Bulliard (1752–93), published in Paris from 1780 onwards. The book's subtitle made it clear that medicinal qualities of plants were still very important, but it also promised to include all the native plants of the kingdom in its fifteen volumes. The illustrations are exquisite, comprising 602 coloured engraved plates, made by Bulliard himself, of which two thirds are of fungi, the artist's focal area of enthusiasm. Bulliard's book shows the beginnings of colour printing: he used the Le Blon-Gauthier method, which entailed a separate printing plate for each colour, for the first time. Bulliard is considered to have been preternaturally skilled in mixing coloured inks by hand, and many of the plates show tiny pin holes, which were used to line up the successive impressions. The extraordinary skill is matched by the delicate aesthetic arrangements of the plants and fungi depicted.

Nonetheless French science was becoming more professional, and it would not be long before such botanical publishing was possible only for those deemed qualified experts by the establishment. In France the Jardin du Roi in Paris became an extraordinary powerhouse under the directorship of Georges-Louis Leclerc de Buffon (1707–88). Buffon's project was entirely Enlightenment in flavour; he had begun his career in mathematics, working in probability theory, but belonged to a Paris circle that included the king's apothecary and his minister for naval rearmament, offering him wide opportunities for advancement.

The Jardin du Roi became the centre of Buffon's energies as he acquired new land to enlarge it and began building display galleries for the expeditions he dispatched globally. The garden became the hub of French botanical activity, soon employing Jean-Baptiste

RIGHT

La Botanique mise à la Portée,
Nicolas Regnault,
(1774–80)

Geneviève was the illustrator for her husband's text, and produced lively originals for engraving, such as this early and accurate image of the growing potato (this page). *La Menthe a Epi* spearmint (facing page, left) and *Le Poireau* the leek (opposite, right).

La Menthe a Epi
Le Poireau ou Porreau
Allium Porrum . Linn . Sp. Pl.
Ital. Porro . Esp. Puerro . Angl. Lecks . Allem. Lauch .

OPPOSITE
Herbier de la France,
Pierre Bulliard,
(1780–93)

Clockwise from top left: Golden henbane, herb of grace, knapweed, agrimony, in copperplate engravings that were some of the first botanical illustrations to be printed in three separate layers of colour.

Lamarck (1744–1829), the French pioneer of evolutionary theory, who took to natural history after being pensioned off injured from the French army. While still a student in 1778 Lamarck published his *Flore Française* (*French Flora*) with delicate black and white drawings. The book's greater significance lies in its pioneering of a method now central to botanical identification – the dichotomous key. The reader is asked a question about the plant they want to identify, for which there are possible answers, each leading them in a different direction. Today most floras are constructed this way, with students encouraged to use illustrations to confirm an identity only after achieving an answer from the key.

5 German botany, Kölreuter and the study of plant fertilization

Elsewhere in Europe, Tübingen in Germany was becoming an important centre for the study of plants. The botanical garden there had originally been created by Leonhart Fuchs (see also page 66), and from 1751 the director was noted field botanist Johann Gmelin (1709–55) who had explored Kamchatka, Russia, under Vitus Bering (1681–1741), who gave his name to the straits. The second expedition, in which Gmelin participated, was dogged by frustrations including a fire that destroyed all of Gmelin's collections and part of his library. Nonetheless he published his *Flora Sibirica* (*Flora of Siberia*, 1747–69), which included more than a thousand species and almost 300 illustrations. An important insight from this journey was Gmelin's modification of his own creationism – his experience of different botanical habitats left him still accepting the religious account, but arguing it had taken place in multiple centres. Travel greatly inspired naturalists to think about why species vary so widely across the earth.

Gmelin's most famous student was Joseph Kölreuter (1733–1806) the son of an apothecary, who studied medicine at Tübingen under Gmelin, and then in St Petersburg, Russia. Despite Kölreuter's religious orthodoxy, he thought carefully and methodically about how fertilization and development of the embryo plant actually took place, and also about how the male and female parents contributed to the offspring. He was first struck while crossing two nicotiana species with each other to produce surprisingly healthy offspring, a phenomenon we now term hybrid vigour. However, when he then crossed the hybrids with each other, no seed was produced. Kölreuter went on to repeat this process with hundreds of other plants, examining the pollen of the sterile third generation under a microscope, finding them to be shrunken husks rather than healthy shapes. He saw this process as a species protecting itself against diminishment, and corresponded with Linnaeus about it.

Kölreuter undertook an extensive programme of experiments encompassing a large range of species. His work shows some of the earliest use of two significant experimental techniques now central to scientific research: repetition and control groups. Kölreuter was careful and methodical, demonstrating that male material in the form of pollen was essential to fertilization and seed formation, and also that both parents contributed equally to the next generation of offspring. He thus undermined a persistent Aristotelian belief that had endured to his day that the father shaped much of the character of a child, the mother acting more as a 'substrate'; as well as rejecting a more recent 'ovist' tenet that the maternal contribution was more important in determining the offspring.

Another student of pollination, Christian Sprengel (1750–1816) worked in rejection

PLANTE SUSPECTE DE LA FRANCE.
LA JUSQUIAME DORÉE.

PLANTE VÉNÉNEUSE DE LA FRANCE.
LA GRATIOLE OFFICINALE.

PLANTE MEDICINALE DE LA FRANCE
LA CENTAURÉE JACÉE ... LA JACÉE DES PRÉS

PLANTE MEDICINALE DE LA FRANCE
L'AIGREMOINE OFFICINALE.

of the then-current belief that visits to flowers by bees and other insects were of no benefit, letting the insects steal nectar, which then sapped the plant's ability to produce seed. (Sprengel would not have his work correctly recognized as significant until Charles Darwin began his own work on the fertilization of orchids, see also page 197). In Sprengel's book *Das entdeckte Geheimnis der Natur im Bau und in der Befruchtung der Blumen* (*The Secret of Nature Discovered in the Construction and Fertilization of Flowers*, Berlin, 1793) he set down his belief that flowers were mostly structures to attract pollinators, understood that nectar was a reward by the plant for entering, and even observed structures in the flower intended to prevent self-pollination. Sprengel noticed that some plants lacked nectar, and differentiated between those that were wind-pollinated and those, like orchids, that 'cheat' their visitors. He was also the first to put in print the suggestion that beehives be placed near crop fields to ensure pollination.

6 The artists of the Austrian archduchy and their travels across the globe

The botanist Nikolaus von Jacquin (1727–1817) first studied medicine and botany at Leiden, Netherlands. In the 1750s he travelled to the Caribbean to collect for the Holy Roman Emperor, Francis I, and by 1768 he was Professor of Botany in Vienna, Austria. Jacquin was highly prolific, early on publishing his *Selectarum stirpium americanarum* (*Of Selected American Plant Races*, 1763), an account of his American finds with nearly 200 plates drawn by Jacquin himself. However, the engravings were not wholly satisfactory – despite an intriguing title page showing indigenous people apparently holding a cornucopia of Caribbean fruits and flowers in offering, as if to the European reader. Among the plants named for the first time in this book was the oil palm, now so problematic in monoculture almost everywhere it is grown.

In response to the perceived drawbacks of his American book, Jacquin began to assemble a team of artists who would bring his treasured discoveries to life. Plates from the *Hortus botanicus vindobonensis* (*A Guide to the University Gardens of Vienna*), appeared in 1770–6, with 300 hand-coloured engravings on copper by the meticulous Franz von Scheidel (1731–1801). The book paid tribute to Jacquin's hard work reconstructing the botanic garden from the poor state in which he found it. Even a spiny thistle, *Onopordum illyricum*, becomes a work of art in the hands of Scheidel. Next, Jacquin's *Florae Austriacae* (*Flora of Austria*) of 1773–8 in five volumes can stand alongside other national works of the age, offering 450 large plates of the region's natural flora. Despite the less exotic nature of the plants illustrated, the artists conveyed beautifully the character of the tiny mountain gentian, the delicate Alpine clematis and the robust service tree.

However, possibly Jacquin's most spectacular achievement came in 1797–1804 with the publication of *Plantarum rariorum* (*Rare Plants*) comprising portraits of many of the rarest plants grown in the imperial garden at Schönbrunn. Amaryllises, bromeliads, palms and aroids fan from the pages in tropical profusion. Such a book speaks of the marshaling power of the European states to collect and transfer species and resources as they saw fit, glorifying the process even as it arguably stripped the new worlds it gazed upon.

Jacquin was an accomplished organizer of talent. Later in life he took under his wing Austrian artist Ferdinand Bauer (1760–1826), who had been born into a court painter's

ABOVE

Selectarum stirpium americanarum,
Nikolaus von Jacquin,
(1763)

The illustrations to this book include this *Hirtella americana*, the American Bastard Cedar, a useful small tree these days often used for revegetating landscapes.

family, but due to poverty was placed early on with his brother Franz (1758–1840) under the care of a priest who gave the boys an artistic training. In 1780 Ferdinand and Franz were despatched to Vienna Botanic Garden to work for Jacquin.

The young men were trained by Jacquin in Linnaean taxonomy, after which the older man began to recommend them for employment. Probably Ferdinand Bauer's most historically celebrated work was on the *Flora Graeca* (*Flora of Greece*, 1806–40), which he made after accompanying the Oxford Professor John Sibthorp (1758–96) on a plant collecting trip to Greece in 1786. The two men met originally in Vienna, where rather poetically they had both gone to see the *Codex vindobonensis* in the flesh in Vienna (see also page 26). The nine-month Greece trip eventually resulted in a book that eventually appeared with almost a thousand hand-coloured illustrations, although it would take the intervention of the Linnean Society's James Edward Smith (see page 53) and the then Professor of Botany at University College, London, John Lindley (1799–1865), to see the book's volumes through final publication.

In his next work *Illustrationes florae Novae Hollandiae* (*Illustrations of the Flowers of New Holland*, 1806–13) Ferdinand Bauer would go even further to ensure botanical

ABOVE

***Flora Graeca*,**
John Sibthorp, illustrated by Ferdinand Bauer, (1806–40)

This extraordinary book would eventually have over a thousand hand-coloured illustrations made after a collecting trip to Greece itself.

RIGHT

***Flora Graeca*,**
John Sibthorp, illustrated by Ferdinand Bauer, (1806–40)

Digitalis ferruginea, the rusty foxglove, and (far right) *Asphodelus ramosus*, both Mediterranean natives, showing the extremely high quality of the images produced by Bauer.

331.
a
Asphodelus ramosus.

ABOVE

***Illustrationes florae Novae Hollandiae (Illustrations of the Flowers of New Holland)*, Ferdinand Bauer, (1806–13)**

Banksia coccinea, by Ferdinand Bauer, showing the delicate, detailed sectional drawings which made his work so scientifically valuable.

accuracy, spending five years from 1805 preparing all the engravings himself. Bauer had accompanied Captain Matthew Flinders (1774–1814), from 1801–03 the British admiralty's choice to lead the first detailed survey of Australia, or New Holland as it was then called. Bauer was an exemplary recorder of what he saw, devising a special numerical system to remember paint colours that he had left behind for weight reasons.

The plates showing *Banksia coccinea* from *Illustrationes* show the enormous delicacy and richness that Bauer brought to his work. As well as the most vividly drawn flower, Bauer also included on his sheet the flower's cross-section, so that the attachments of the individual elements could be seen, zoomed-in details of the petals, the cone after the flower had fallen away, and deeply attentive depictions of the seeds and seed cases as they would disperse in the wild. Nonetheless, the huge publishing enterprise was a financial failure.

Bauer's brother Franz had a more straightforward career, where on arrival in London he was set up by Joseph Banks as the first full-time botanical illustrator at Kew Gardens, on a salary of £300 a year. His court childhood had perhaps prepared him well for living side-by-side with the royal family at the Gardens, and he was occasionally called to tutor the royal children in painting. In 1796 Franz published the wonderfully titled *Delineations of Exotick Plants, Drawn and coloured and the Botanical characters displayed according to the Linnean System*, showing to what extent Linnaean botany had become an appealing 'selling point' to likely audiences. Franz Bauer is buried in the parish church on Kew Green, next to Kew Gardens to the west of London.

7 The widening of the colonial grip by Britain, France and Spain: from Bougainville to Darwin

Collection and publishing were not simple expressions of curiosity and they often formed part of missions to stake territorial claims on parts of the world. The publication of journals, specimens and reports from such voyages was an important way of asserting science as a 'pure' pursuit, not entangled as it actually was with political and national ambitions. The pattern of these voyages perhaps began with Captain James Cook (see page 45) and his French contemporary, Louis de Bougainville (1729–1811).

Bougainville had become famous for completing the first scientific circumnavigation of the globe in 1763, meaning that there were naturalists and mapmakers onboard. Bougainville's account of this voyage, *Voyage autour du monde* (Voyage Around the World) was a contemporary bestseller from 1766 partly for its accounts of Tahitian life, but it did not mention what has latterly become one of the most interesting parts of the story. Jeanne

Baret boarded the voyage posing as botanist Philibert Commerçon's (1727–73) male valet and botanical assistant from 1766–69. Baret and Commerçon did much significant scientific work on the voyage, but there is debate about the point at which she was revealed to be his partner.

Commerçon died mid-travels in Mauritius, never returning to France, but his many collections – made in collaboration and sometimes wholly by Baret – rest now in the Museum of Natural History in Paris, on some 1,735 specimen sheets. On her return, Jeanne Baret's contribution was recognized by the French government, who awarded her a life pension from a fund for invalided servicemen. On 27 July 2020, the internet marked her 280th birthday in one of its greatest tributes, the day's Google Doodle. Entwined around her is the plant she and Commerçon collected, and later named after the expedition's leader: Bougainvillea.

Bougainville's accounts of the Tahitians in particular were significant in encouraging a view of the islanders as 'simple innocents' living in a kind of paradise. Denis Diderot (1713–84) after finishing work on his *Encyclopédie* (*Encyclopedia*, 1751–1772), wrote his *Supplément au voyage de Bougainville* (*Supplement to the Voyage of Bougainville*, 1796) espousing just such a view. However, most European observers tended to find non-European lifestyles primitive and unprogressive compared to their own, helping to justify appropriation and control. In some ways, the naturalists' gaze was extended to indigenous people, as if they were part of a continuum of new world resources to be surveyed from plants through to human beings, a view expressed in actions from slave-ownership to metropolitan exhibitions of living native people.

As the nineteenth century began, these imperatives became sharper. The voyage of the *Beagle* from 1831–36, for example, including young naturalist Charles Darwin (1809–82), was part of a British Royal Navy mission ostensibly to survey the coasts of South America, but it also provided the navy with a soft excuse for testing the defences of the Spanish Empire and locating trading partners who might be willing to build relationships with the British. Another such voyage was that of Darwin's best friend Joseph Hooker (1814–79) on the *Erebus*, eventually published as the *Flora Antarctica* (1844–47), illustrated by Walter Hood Fitch (1817–92). Hooker travelled on this voyage not as a naturalist – there were not sufficient funds available – but as a naval surgeon, something of a disappointment to him.

It is clear that botanical collectors and explorers from Europe imagined huge benefits flowing from their industry, as the manuscript volumes of the Royal Botanical Expedition to New Granada (1783–1816) show. Its mastermind, José Celestino Mutis (1732–1808) was a priest and physician who had campaigned tirelessly for twenty years asking the Spanish king to send a botanical study to New Granada, the name of the Spanish Empire in South America. Mutis was a meticulous

BELOW

José Celestino Mutis, by R. Cristobal, (1930)

Mutis, a priest and medic active in New Granada, oversaw possibly the largest colonial collecting expedition to date, though it was so voluminous it was never properly processed.

and ordered man, determined to record not only plant species but also the use of woods and other economically useful products. Yet sometimes information was collected beyond what could be processed. About 30,000 plant specimens were returned by Mutis to the Royal Botanic Garden in Madrid, but little heed was paid to these arrivals in the capital. The plates made by the eight or so expedition artists are today recognized, though, for painters far outnumbered actual botanists, Mutis having understood that the impact of the discoveries in Madrid would be the real test. However, the expedition's work was not published until 1954, running into more than sixty volumes, and remains unfinished.

Nonetheless, some botanical travellers hoped to transcend national boundaries. One of these was Alexander von Humboldt (1769–1859), the holder of a holistic vision of the Earth and its living environment. Authorized by the Spanish King Charles IV to travel to South America, Humboldt observed the changes wrought on the environment there by the farming of export crops such as sugar, cacao and cotton. He theorized that the widespread felling of trees had made soils drier and yields lower, becoming arguably the

BELOW

Essai sur la géographie des plantes (Essay on the Geography of Plants), **Aimé Bonpland and Alexander von Humboldt, (1805)**

This physical diagram of Chimborazo in the Andes and the surrounding lands shows the plants growing at each level of habitat, and is one of the first diagrams to show this kind of spatial separation of species.

first to write of human-made climate change. On arriving in Bogotá, Colombia, in 1801 he met with Mutis himself, who gave him generous assistance, and Humboldt was struck by the quantity of artists in Mutis' employ.

Humboldt, though, travelled on, measuring and recording as he went; eventually he would publish accounts vividly depicting the South American natural world, which would form the foundations of later environmental science. Humboldt often presented his massed data in the form of graphs and diagrams, writing that they 'speak to the senses without tiring the intellect', enabling the scientist to present a large number of facts at once. He was an innovator in recognizing the vegetational zones and other natural patterns, such as temperature; Humboldt is today noted for having pioneered these methods of recording and displaying patterns in scientific data.

Such Western scientific approaches were spreading beyond the realms of purely European influence. The Japanese botanist and pharmacologist Ono Ranzan (1729–1810) began his career by studying traditional Chinese medicine, but quickly became interested in the Western *materia medica*, translating the herbal *Cruydeboeck* of Rembert Dodoens (1517–85) into Japanese. Ono travelled to collect Japanese herbal medical remedies for his important work, *Honzō kōmoku keimō*, (*Dictated Compendium of Materia Medica*, 1803–06); he firmly advocated that Western scientific practices, such as experimentation, should supplement traditional medicine. In distancing itself from the Chinese classical medical tradition, the book also represented the Japanese desire to elude or even surpass China. National identity would be an increasingly significant aspect of botanical practice in the century to come.

8 The indigenous eye looks back, from British India to the Caribbean

European artists were enormously skilled in their own working practices, but in many cases a local artist was the best choice for recording the flora of their own area. Native artists were used by the British East India Company for illustrations – exquisite drawings including plants, birds and insects, depicting them together in a kind of ecological unity. Traditionally these artists were simply identified as part of the 'Company School' (see also page 250), and yet recent scholarship has argued that individual artists can be identified, and that the period demonstrates a very clear new fusion of styles, with the detailed colourism of the Indian artists coming in to work alongside the Western botanical requirements and making something completely new. The paintings are distinguishable both by the painters, and also by geographical origin, as each region tended to have a local style.

Botanic gardens throughout the expanding British Empire were part of an informal network centred on Kew Gardens, in recognition of the importance of plant-based commodities in the imperial economy. Nathaniel Wallich (1786–1854) was a surgeon-botanist from Denmark who helped to establish the Calcutta Botanic Garden where he worked from 1817 until his death, working on his own projects but also hosting and assisting many European plant collectors. He compiled a herbarium including some 20,000 names, and his work is recognized today in a number of plants named 'wallichiana'. *Plantae Asiaticae rariores* (*Rare Asian Plants*, 1829–32) was Wallich's effort to

ABOVE

Udagawa Yōan (1798–1846)

Japanese scholar Udagawa Yōan was known for Western science into Japan. He described and established many terms for the morphology, anatomy, physiology and taxonomy of plants, many of which are still used today.

ABOVE

Alexander von Humboldt (1769–1859), by A. Neumann, (1883)

From a wealthy family, Humboldt was able to follow his keen interest in the natural world: travelling, researching and publishing.

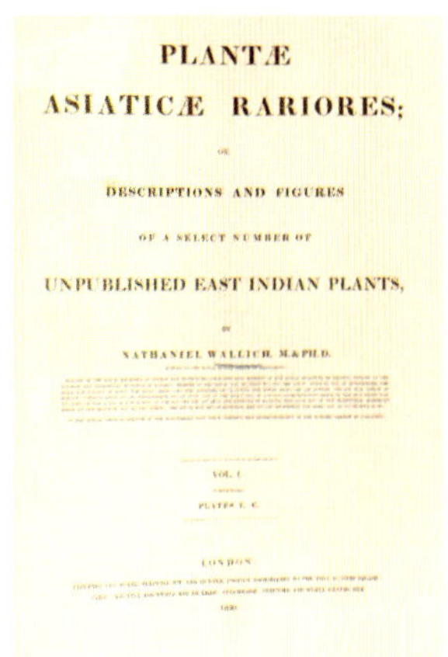

PLANTÆ
ASIATICÆ RARIORES;

OR

DESCRIPTIONS AND FIGURES

OF A SELECT NUMBER OF

UNPUBLISHED EAST INDIAN PLANTS,

BY

NATHANIEL WALLICH, M.&PH.D.

VOL. I

PLATES I. C.

LONDON

OPPOSITE LEFT

Nathaniel Wallich (1786–1854)

Wallich worked tirelessly to expand the botanic garden of Calcutta, hosting European naturalists and sending specimens back to Kew Gardens and others. His most important publication was *Plantae Asiaticae Rariores, (Rare Asian Plants)* 1830–32.

OPPOSITE RIGHT

***Curcuma parviflora,* Nathaniel Wallich, (1834)**

Curcuma parviflora, White Angel, from the ginger family, a native of the Myanmar region, from a journal article by Nathaniel Wallich in 1830.

THIS PAGE

***Rhododendron arboreum,* Nathaniel Wallich, (1834)**

Rhododendron arboreum, introduced from Nepal by Wallich. This gorgeous hand-coloured engraving comes from the young Joseph Paxton's brand-new *Magazine of Botany*, 1834.

ABOVE

Paul Kitaibel,
(1757–1817)

Kitaibel was Professor of Botany at the University of Pest, and prepared a spectacular flora from 1802–12 glorifying Hungary's natural riches at a time of growing nationalism.

RIGHT

Aylmer Bourke Lambert,
(1761–1841)

One of the founding members of the Linnean Society, he shaped the early history of British botany.

LEFT

Pinus palustris,
Aylmer Bourke Lambert,
(1761–1842)

Pinus palustris, the American longleaf pine, from Aylmer Bourke Lambert (1761–1842)'s *Descriptions of the genus Pinus*, one of the first studies of a single botanical genus.

THIS PAGE

Pomologia, (the study of apples and pears)

Pomologia, (the study of apples and pears), was written by Johann Hermann Knopp in 1758, identifying over a hundred varieties of apple. It began the modern scientific study of fruit growing.

OPPOSITE

A Description of the genus Roman

Pinus longifolia, from the Himalayas, in an elegant engraving from Lambert's *A Description of the genus Roman*. Lambert was an energetic networker and knowledge collector, acquiring specimens and whole herbaria from other experts. This pine is now known as Roxburghii.

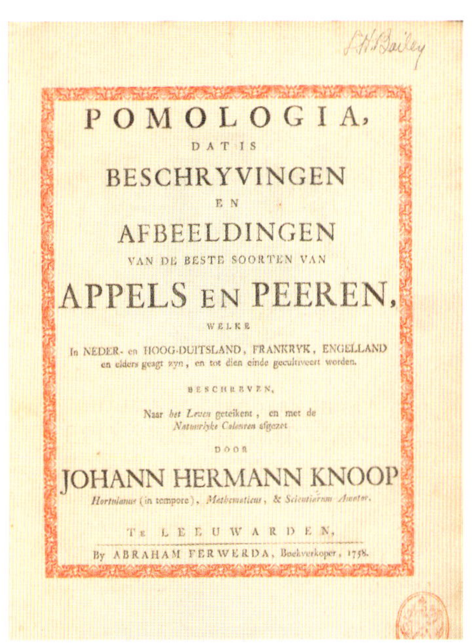

POMOLOGIA,
DAT IS
BESCHRYVINGEN
EN
AFBEELDINGEN
VAN DE BESTE SOORTEN VAN
APPELS EN PEEREN,
WELKE
In NEDER- en HOOG-DUITSLAND, FRANKRYK, ENGELLAND en elders geagt zyn, en tot dien einde gecultiveert worden.
BESCHREVEN,
Naar *het Leven* geteikent, en met de *Natuurlyke Colouren* afgezet
DOOR
JOHANN HERMANN KNOOP
Hortulanus (in tempore), *Mathematicus*, & *Scientiarum Amator*.
TE LEEUWARDEN,
By ABRAHAM FERWERDA, Boekverkoper, 1758.

bring to European attention many previously unfamiliar East Asian plants.

The book came with 294 elegant lithographs by the celebrated London engraver Maxim Gauci (1774–1854), but the original illustrations were made in India by two local artists, Gorachand (active 1830–32) and his colleague Vishnu Prasad (active 1830–32) (see also page 254). Intriguingly, there are also traditions of Chinese artists working in and around the trading ports of South Asia, producing work mainly for Western eyes, especially for East India Company grandees. One of the few of these artists to be named rather than anonymous is Win Achun (active nineteenth century), who made an exquisite curling honeysuckle on rice paper.

Enslaved artists were used by Scottish botanist Alexander Anderson (1748–1811), the superintendent of the British colony of St Vincent's Botanic Garden – the first one to open in the Caribbean – to produce a proposed but unfinished flora of the Caribbean. Anderson's catalogue named the almost 2,000 taxa in his care by their Latin names, but also by their Carib and 'Afro-Caribbean' names, suggesting Anderson had made use of enslaved people's skills and knowledge here too. Scholars working today have increasingly tried to identify the individual artists, even where they remain anonymous, by using stylistic differences in their work. Unusually, John Tyler, an Antiguan free person of colour working for Anderson between 1785–1811, signed his work in immaculate copperplate writing.

BELOW

Pierre-Joseph Redouté, *Peintre des Fleurs,* Claude-Antoine Thory, (1817)

Redouté was famous during his own lifetime for his exceptional flower paintings, which eventually led to his close working relationship with Josephine Bonaparte.

9 Pierre-Joseph Redouté

Scientific illustration was increasingly pragmatic in outlook, but there were still high points of artistic flourish. Pierre-Joseph Redouté (1759–1840) is probably the most famous flower painter of all time, but was born in Belgium into a humble artisan background, his family travelling to paint signage and commercial decoration, later making scenery for new theatres in Paris. It was in pre-Revolutionary Paris that Redouté began to focus on botanical painting, working from living models where at all possible, and despite the foment of the era establishing networks with renowned botanists and horticulturalists.

His most important breakthrough was making the acquaintance of the acclaimed botanist Charles Louis L'Héritier de Brutelle, known just as L'Héritier (1746–1800). L'Héritier commissioned Redouté to make accurate botanical illustrations for his descriptive publications, taking him in the 1780s on a trip to the Kew Gardens, then bringing him back to work at Paris' famous Jardin des Plantes.

Redouté was fascinated by the technical side of printing and carefully studied the technology by which his originals would be reproduced. As he mastered printmaking, which was undergoing significant changes at the time, he embraced an innovation called 'stipple engraving' where very subtle variations in tone were possible, giving rise to extraordinary new colour effects that could be reproduced many times due to the hard-wearing nature of the metal plate. Redouté spanned two worlds, producing exceptional flower paintings for aristocratic patrons as well as scientifically accurate images for the use of natural history.

LEFT

***Les Roses,* Pierre-Joseph Redouté, (1817–24)**

Rosa centifolia, here hand-coloured with sophisticated stipple shading, a technique taken to its greatest possibility in Redouté's work.

THIS PAGE

***Choix Des Plus Belles Fleurs*,**
Pierre-Joseph Redouté,
(1827)

Narcissus gouani, the double daffodil, here in delicate image by Pierre-Joseph Redouté, around 1833.

OPPOSITE

Les Liliacée,
Pierre-Joseph Redouté,
(1802–1816)

The Turk's Cap Lily from *Les Liliacées* by Redouté, a native of North America.

RIGHT

***Choix Des Plus Belles Fleurs*, by Pierre-Joseph Redouté, (1827)**

Plumbago caerulea, a native of South Africa, named by Carl Kunth who had been a botanical assistant to Humboldt, and a bunch of narcissi (far right). *Narcissus tazetta*, the bunch-forming wild daffodil from which many cultivated narcissi are descended.

Narcisses à plusieurs fleurs
Narcissus tazetta
P. J. Redouté

OPPOSITE

Alphonse de Candolle, (1778–1841)

Alphonse de Candolle was arguably the first to describe the global vegetational zones we would now understand to be shaped by environment and climate.

OPPOSITE

***Coleotrype natalensis* (1881)**

Coleotrype natalensis, a South African relation of the Bromeliads and gingers, in a description published in 1881 by de Candolle's son, Alphonse Pyramus.

One of Redouté's most notable achievements is his collaboration with Joséphine Bonaparte, the first wife of Napoleon Bonaparte. Joséphine, an avid lover of botany and roses in particular, employed Redouté as her official court artist. Under her patronage, Redouté created some of his most famous and cherished works, including the collection *Les Roses* (*Roses*), from 1817, which featured meticulously detailed watercolor illustrations of over 170 rose varieties. These illustrations showcased not only his exceptional technical skill but also his ability to capture the delicate beauty of each flower. Regrettably the original paintings for *Les Roses* were not kept together but sold off piecemeal, and their whereabouts today are known only patchily.

10 Botany amid European turbulence

French botany had a strong academic centre in the Jardin des Plantes (as the Jardin du Roi had been renamed after the Revolution, 1789–1799), and several important figures emerged at this time. Michel Adanson (1727–1806) had moved away from Linnaeus' system, which focused only on the sexual parts of the flower. Adanson had undertaken a five-year voyage to Senegal, which was deeply influential on his view of the natural world, and he found the works of Tournefort (see page 111) and John Ray (see page 106) fruitfully useful. In 1763 he published *Familles naturelles des plantes* (*The Natural Families of Plants*), trying to view the plant world as essentially a family tree. His work did not meet with universal acclaim at the time, but its influence on the subsequent generation is measurable.

Next came Antoine Laurent de Jussieu (1748–1836) who admired Adanson and was the first botanist to create a natural classification to include all the flowering plants, accepting that there are close familial relationships between the members of each related group. Jussieu came from a large family of botanists and graduated in medicine in 1770, immediately becoming a botany demonstrator at the Jardin du Roi. His first specialist work was the classification of the Ranunculaceae, the buttercup family, in 1773; but on 4 August 1789, the day when the French Revolution abolished feudal rights and privileges, he published his masterwork the *Genera plantarum*, comprising of a hundred plant families organized (somewhat ironically) hierarchically, into divisions, classes and orders.

Jussieu then had to set botanical work aside and was placed in charge of the Paris hospitals, but on re-establishment of government in 1790, he helped to reorganize the Jardin des Plantes in the city to create the Museum of Natural History. He was delighted to find himself in charge of many new collections appropriated by the revolutionary armies from the aristocracy both in France and abroad. This may have somewhat prejudiced the botanists of other nations against his work, because it could be seen as dangerously radical; however, its value was eventually appreciated.

His most important insight was to view the families of the plant world as differentiated by a range of characters, not by one set. Some might be more important than others; some might be more influenced and changeable under environmental conditions, and thus less useful. His principle was '*pesés, et non comptés*' – 'weighed, not counted' – indicating that more judgment was to be used. The implication of this was that Linnaeus' system of classification, which could be used by anyone able to count, was to be displaced by one where only experts could truly be trusted to do the work. Today, seventy-six of Jussieu's families remain in use, as opposed to just eleven of Linnaeus'. His work can be seen as the beginning of botany as a science, fusing the work of naturalists, physicians, herbalists and philosophers.

11 Alphonse de Candolle

The Swiss naturalist Alphonse de Candolle (1778–1841) was born in Geneva, studied under the pastor botanist Jean Pierre Vaucher, and quickly found a career for himself in a private herbarium. His skill in classification was quickly noted in the capital by Cuvier and Lamarck and he was given charge of the 1805–15 project of the French government to compile a complete *Flore Française* (*French Flora*). Like Jussieu, he advocated a natural classification, setting out to write a complete treatise according to his system. Despite finishing ten volumes of the *Prodromus systematis naturalis regni vegetabilis* (*The Natural System of the Vegetable Kingdom*, 1824–39) before his death in 1841, and his son's efforts to complete the work, the series remained unfinished, being arguably too ambitious.

de Candolle did not subscribe to evolution, but he recognized in the world's flora such a variety that he argued for twenty botanical regions, each separate zones of creation. Despite this creationist view, he began to be keenly interested in what he called 'disjunct species', belonging to apparently the same species, yet in the present day so geographically separated as to raise questions about their past history. Later in life, he acknowledged that perhaps the 'deviation of anterior forms' was the most likely explanation for what he had seen in the natural world; otherwise known as evolution. Among his other accomplishments, de Candolle invented the word 'taxonomy', and was the first to propose the idea of 'sub-species' to help biologists account for where species definitions do not quite work. As a result, his ideas were adopted at one of the first international biological congresses in the 1870s.

12 The scientific establishment in London and the world of William Hooker

William Jackson Hooker (1785–1865) was one of Britain's most significant botanical minds. *Muscologia Britannica* (Longman, 1827) is a book about the mosses of the British Isles, by Hooker and his colleague Thomas Taylor (1786–1848), an enthusiastic botanical physician, who together drew the illustrations and supervised all the engraving. Mosses and liverworts were a childhood passion for Hooker, and his early botanical expertise meant he was elected in 1806 to London's Linnean Society while still working as the manager of a brewery. After a fortunate inheritance, Hooker was handpicked by an elderly Joseph Banks for a research trip to Iceland in 1809, and Banks then supported

Hooker's appointment in 1820 as Professor of Botany in Glasgow, where he wrote and published this book.

Hooker epitomized the new professional spirit in science of the early nineteenth century, as governments and universities across the globe began to recognize the value of formally supporting scientific work and new institutional posts were created, enabling new qualifications to be taught and awarded. Amateurs would still have access to science but now there would be university-educated gatekeepers. In 1821 Hooker published his *Flora Scotia* (*Scottish Flora*), which went through many editions. It is significant that despite the perceived greater accuracy of the new natural classification proposed by Jussieu, Hooker continued to use the Linnaean system in this book for popular use, believing it to be more easily masterable by amateur botanizers. Hooker also included his own classification, believing it reflected genuinely-related plant families, more closely descended from one another than from other less similar ones. In fact, the full title made it clear that the plants were arranged 'both according to the Artificial and the Natural Methods', to give readers both possibilities. The days of natural history were numbered, as such judgments were to be made in future on their behalf by educated experts. From now on, it would be possible for a growing number of such experts, almost all men, to call themselves 'scientists'.

BELOW

Sir William Jackson Hooker, (1785–1865)

Hooker in a photograph taken in later life, when he continued to work with great industry as director of Kew Gardens.

Muscologia Britannica;

CONTAINING

THE MOSSES

OF

Great Britain & Ireland,

SYSTEMATICALLY ARRANGED AND DESCRIBED;

WITH PLATES ILLUSTRATIVE OF THE CHARACTERS OF THE GENERA AND SPECIES.

BY

WILLIAM JACKSON HOOKER, F.R.S. A.S. L.S.

AND MEMBER OF THE WERNERIAN SOCIETY OF EDINBURGH,

AND

THOMAS TAYLOR, M.D. M.R.I.A. & F.L.S.

AND FELLOW OF THE KING AND QUEEN'S COLLEGE OF PHYSICIANS OF IRELAND.

LONDON:

PRINTED BY RICHARD AND ARTHUR TAYLOR, SHOE LANE;

FOR LONGMAN, HURST, REES, ORME, AND BROWN, PATERNOSTER ROW.

1818.

ABOVE

Muscologia Britannica, William Jackson Hooker and Thomas Taylor, (1818)

Hooker was not just an enthusiast for the complex and flowering plants; he was an expert on mosses and liverworts, publishing *Muscologia Britannica* in 1818 with his friend Thomas Taylor.

LEFT

Curtis' Botanical Magazine (1826)

William Hooker became editor of *Curtis' Botanical Magazine* in 1826, and this is typical of the delicate illustrations he commissioned.

RIGHT

Plantarum (Images of Plants),
William Hooker
(1837–54)

Cucurbita pepo, the pumpkin, from William Hooker's *Icones Plantarum* (*Images of Plants*), a highly illustrated series of books published from Kew herbarium specimens.

OVERLEAF

Exotic Flora
William Hooker
(1823–27)

Goodyera procera, the bottlebrush orchid, left, and *Synedrella nodiflora*, the Cinderella Weed, right, in this handcoloured copperplate engraving by J. Swan, after a botanical illustration by William Hooker.

Tab. 148.

39
1
2
3
4
Goodyera procera
J. Swan Sculp.t

60
1
2
3
4
5
Synedrella nodiflora
J. Swan Sculp.t

BOTANY BECOMES A SCIENCE
(1830–1950)

Botanists now increasingly sought to understand how plants grew and nourished themselves. As early as the seventeenth century, experimenters had already proved that a gas that sustained animals was produced by plants in the presence of sunlight; in the early 1800s Nicolas de Saussure (1767–1845) in Geneva confirmed that plants required light, water and carbon dioxide to grow. In the 1850s, French mining engineer Jean Baptiste Boussingault (1801–87) grew plants without organic matter but with access to nitrogen compounds, showing that their growth was in proportion to the nitrogen available. Botanists now realized plants needed nitrates and phosphates for full growth, which could be added as fertilizers.

The German scientist Justus von Liebig (1803–73) set clear aims in books such as *Die organische Chemie in ihrer Anwendung auf Agricultur und Physiologie* (*Organic Chemistry in its Application to Agriculture and Physiology*, 1840) to use chemistry to increase food production in agriculture. He showed that plant growth was limited by the resource of which they had the least: whether that be light, water, carbon dioxide, nitrogen or trace elements. Liebig innovated further by organizing biological laboratories in teams, allowing several people to work on the same scientific problem.

Cell biology became a key area of research, with better imaging via microscopes and new analysis techniques facilitating a greater understanding of the cell's interior workings. In 1837 in Tübingen, Germany, Hugo von Mohl (1805–72) observed cells dividing for the first time and described the chloroplasts, the organelles later understood to be concerned with photosynthesis.

While identifying the cell's main structures, scientists also began to try to understand its chemistry. In 1877 the German physiologist Wilhelm Kühne (1837–1900) first used the term 'enzyme' to describe the substances that catalyzed reactions in the living world, and in 1894 Emil Fischer (1852–1919) suggested the lock-and-key model for how they act. Later they used 'tracer' molecules to track what was happening in key cellular reactions, and by the early twentieth century looked to radioactive molecules to act as markers. Cornelis Van Niel (1897–1985) made significant discoveries about photosynthesis by studying bacteria. And Robin Hill (1899–1991) used chloroplasts isolated from the living cell to show that they would still give off oxygen in the presence of light.

Botanical taxonomy and the classification of new species continued to be extremely important; their spread to other parts of the world measured the eventual global acceptance of Linnaean systematics. In the middle of the nineteenth century, significant botanical discoveries were being made by naturalists like Charles Darwin (see also page 196)who had little formal training and whose career was enabled by a substantial private income. But from the later nineteenth century onwards, the most important innovations in botany increasingly took place in laboratories, run by salaried scientists. Botany was rebranded 'phytology', or plant science, with practitioners distancing themselves from their gentler predecessors. Luxurious publishing of large-scale illustrations continued, but became less central to taxonomy and perhaps more important to horticulturalists, presenting the rarest of new market introductions to gardeners. In Japan, the local illustrative style fed into presentations of regional flora, as seen in Iwasaki Tsunemasa's *Honzō Zufu*, a marriage of Western taxonomy with Japanese aesthetic traditions.

1 Botany in Japan

Honzō Zufu (*Botanical Atlas*), an extensive and astonishing account of Japanese flowering plants, was published from 1828 onwards, eventually running to ninety-three volumes. Its author, Iwasaki Tsunemasa (1786–1842), came from a samurai family in the fortified town of Edo (modern-day Tokyo). Iwasaki became superintendent of the botanical gardens of the Shogun (the local ruler), taking collecting trips for native Japanese plants across the length of Honshu Island, and in 1818 published his first book, *Sōmoku sodate-gusa* (*Growing Trees and Plants*).

In 1823, the German botanist Philipp von Siebold (1796–1866) came to live in Japan, keen to collect Japanese plants and employing local artists such as Kawahara Keiga (1786–*c.* 1860) to create botanical illustrations. Siebold's collections were returned to Europe, where Siebold eventually published his *Flora Japonica* (*Flora of Japan*) in cooperation with the botanist Joseph Gerhard Zuccarini (1797–1848). As a local botanist, Iwasaki was keen to meet with Siebold and discuss the classification of Japanese endemics. However Siebold's collections would have a long-term impact on Japanese botany: in a pattern also echoed elsewhere across the globe, the 'type specimens' of many Japanese plants – the single specimen taken to define a species – would end up held in European collections.

BELOW

***Honzō Zufu*, Iwasaki Tsunemasa, (1828)**

This lovely woodprint book would originally have been bound in a typical Edo East Asian style, called fukuro-toji, printed only on one side and sewn together.

THIS PAGE AND OPPOSITE

Honzō Zufu,
Iwasaki Tsunemasa,
(1828)

Camellia images from Iwasaki's authoritative work, one of the first to combine Linnean botany with Japanese knowledge of native plants.

本草圖譜
巻之八十八
ひらきつばき
三
一種
ちりつばき
花の形やまつばきに似て
紅色く散る時
辮離れて落つ

RIGHT AND OPPOSITE

***Album of 100 Flowers*, Hisui Sugiura (1920–22)**

Aconitum uncinatum (left) and *Alium fistlosum* (Welsh Onion), from Hisui's Album of 100 Flowers (1920–22). Japanese illustrators of the *fin de siecle*, such as Hisui Sugiura, tended to work in this realistic yet sparsely elegant style.

ABOVE

Tomitaro Makino. (1862–1957)

Tomitaro Makino another central figure in Japanese botany, produced a voluminous guide to the islands' flora in 1940, still known as 'Makino's'.

Iwasaki had trained with expert predecessors, in particular Ono Ranzan (1729–1810), but felt strongly that the texts of Japanese botany were being hampered by their lack of illustrations. Ono's work, for example, had been organized according to the sixteenth century system laid out in the Chinese pharmacopeia the *Ben Cao Gang Mu* (see page 126), itself originally unillustrated. In his foreword to the *Honzō Zufu*, Iwasaki set out his intention to address this, arguing that words and pictures complemented each other to aid identification.

In particular, Iwasaki believed in the importance of colour. His work differed from the books contemporaneously appearing in Japanese, influenced by the new Linnaean botany, which focused on structures, especially fruit and flowers. Iwasaki did not choose to enforce blanket standards for visual images, preferring instead to let images spread like the flowers themselves, often across two pages at once, choosing to depict what gave the most poignant sense of a plant. This stylistic choice has made for many instantly recognizable 'plant portraits'.

2 Looking down the microscope

Early microscopy was hampered by technical issues, especially with the manufacture of lenses, but Charles Mirbel (1776–1854) in his *Traité d'anatomie et de physiologie végétale* (*Treatise on Plant Anatomy and Physiology*, 1802) observed that all plant cells were surrounded by a continuous membrane. The 1809 edition of his book came with nine lithographed plates bursting with detail, showing sections through a plant's structure; his contemporary Johann Moldenhawer (1766–1827) also distinguished some basic plant anatomy, identifying vascular tissues and stomata.

However, until about 1830, blurred images put scientists off using microscopes. How could they be sure what they had actually seen if the image itself was suspect? Also there were issues where imperfect lenses focused different colours of light at slightly different distances, giving the final image a set of distinctive halos. However Joseph Jackson Lister (1786–1869) became convinced that microscopes could be improved by changing the way they were assembled, and he enlisted the instrument-making firm of William Tulley. These new microscopes soon achieved popularity, for example letting Lister's friend Thomas Hodgkin identify among red blood cells Hodgkin's Disease.

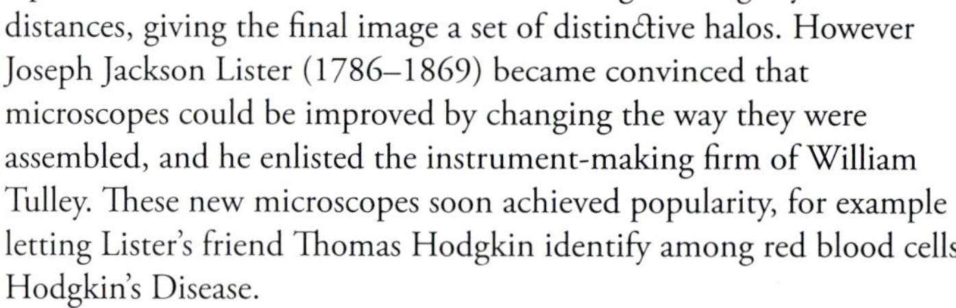

Manufacturers responded quickly to the new demand, and in laboratories from Britain and France to Austria and Germany a microscope became seen as standard equipment. In 1831 the Scottish botanist Robert Brown (1773–1858) read an account to the Linnean Society of using a microscope to see the cell nucleus for the first time. Scientists began to appreciate the significant differences between animal and plant cells, and they understood for the first time that a structural cell wall only occurs in plants. In the late 1830s, the German botanist Matthias Schleiden (1804–81) and zoologist Theodor Schwann (1810–82) theorized that cells were the basic building blocks of life, and published beautiful images made by microscope in Schwann's *Mikroskopie* (*Microscopy*, 1839) and Schleiden's *Beiträge zur Phytogenesis* (*Contributions to Phytogenesis*, 1838).

Schleiden also encouraged the young Carl Zeiss (1816–1888) to start his firm specializing in fine optics, in particular by placing a significant number of orders. Schleiden argued that every part of a plant was cellular, including seeds, fruits and nuts – an insight that had never been made before. The challenge that then followed was to explain what happened at fertilization, if each parent was to contribute half of what the offspring inherited.

The recognition that all life was based on the cell allowed biology to be seen as a single field for the first time. But the practice of science itself was changing. The gentlemen amateurs and well-educated upper-class women of the early nineteenth century were being slowly supplanted by those who had received specialist educations and degrees in the relevant subjects, and their contributions were no longer as welcome. These university men published papers and wrote textbooks. This process was not immediate, and was more noticeable earlier in Germany, for example, than in England, but eventually it would determine the whole shape of botany.

OPPOSITE BELOW

Matthias Schleiden, unknown artist, (*c.*1855)

Schleiden was instrumental in bringing the microscope into common scientific use, allowing biologists to study life at the cellular level for the first time.

3 Orchidmania

Orchids were one of the most fashionably collectable plants of the nineteenth century. Their extraordinary variations on their basic floral plan, with frills, patterns and colour, seem to have aroused an excitement that is hard to explain in purely botanical terms. It seems likely that the vast cost of orchid rarities contributed. Publishing soon cottoned on to this frenzy, as seen in *Illustrations of Orchidaceous Plants*, 1830–38, the collaboration between artist Franz Bauer (see page 160) and the English botanist and horticulturalist John Lindley. The two were most likely introduced by Joseph Banks, perhaps at Kew Gardens, where Bauer was the first artist-in-residence. Recent technical improvements in glass-making, heating and pest control meant that the fates of tropical orchids grown in London had changed for the better, so that these two enthusiasts could join together in celebration. Lindley had begun as Robert Brown's assistant, at Banks' own herbarium, but from 1822 he was the Assistant Secretary of the Royal Horticultural Society and later the first Professor of Botany at the new University College London.

Bauer had begun using the novel microscope technology to examine the very small reproductive details of orchids, in order to elucidate their taxonomy, and Lindley wrote in his preface of how it had been deemed essential to shift away from copperplate engravings to represent these. The book used the more expensive and detailed lithographs, some shown at magnifications of up to 400 times. Lindley pioneered classifying the orchids by looking at the arrangement of pollen on the column, a technique still in use today.

One of the most significant orchid books was produced by James Bateman (1811–97) the landowner who made the extraordinary garden at Biddulph Grange in Staffordshire. *The Orchidaceae of Mexico*

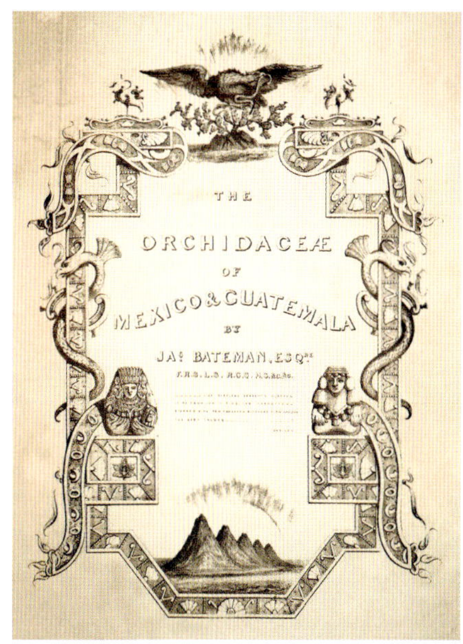

BELOW

***The Orchidaceae of Mexico and Guatemala*, James Bateman, (*c.*1845)**

This inviting title page only hints at the richly illustrated volume to come, one of the largest botanical books ever produced.

THIS PAGE AND OPPOSITE

The Orchidaceae of Mexico and Guatemala, James Bateman, (*c.*1845)

Epidendrum erubescens, now *Artomira erubescens*, from Mexico, and *Galeandra baueri*, named for Bauer, from the Amazon basin.

GALEANDRA BAUERI.

and Guatemala (1837–43) measured 75cm (30in) tall and weighed in at over 17kg (37lb), one of the very biggest botanical books ever produced.

The illustrations were drawn by Sarah Anne Drake (1803–57), Augusta Withers (*c.* 1793–*c.* 1864) and others, and show the spectacular botanical richness of the orchid family. The book includes the first published illustrations for about 90 per cent of the plants shown in the plates. The lithography was an extremely specialist technique, requiring hand-colouring and the subsequent addition of layers of varnish to deepen the tonal qualities of the images. Only 125 copies were ever made, and the book now sells for up to US$100,000 at auction. It was an extremely luxurious piece of publication and reflects the great lengths to which orchid collectors were prepared to go for completism.

4 Darwin and plants' place in evolutionary theory

The most important shift for nineteenth century botany was the coming of evolutionary theory. Once evolutionary ideas had been accepted, the relationships between different families of plants could be seen as genuinely familial, but also functional anatomy and physiology of plants could become the study of how these adaptations helped plants survive. Although other evolutionary theories had been presented to the public, it would be English naturalist Charles Darwin (1809–82) who laid the scientific basis for the wide acceptance of evolution.

Darwin's interest in plants began at university in Cambridge when he was taught botany by John Stevens Henslow (1796–1861). Henslow taught Darwin how to collect, describe, annotate and use a microscope to see further detail, and it was he who recommended Darwin for the naturalist's position on the *Beagle*.

Darwin's account of *The Voyage of the Beagle* was published to great interest in 1839, and his tales of riding in the pampas and seeing gigantic Galapagos turtles are both interesting and charming. However, his private thoughts had turned to more serious matters: the nature of a species itself. How did there come to be many species of closely related plants or animals on different islands in a chain, each subtly different? Could the explanation lie in species themselves being mutable?

BELOW

Charles Darwin, photographed by Julia Margaret Cameron, (1868)

Darwin visited Cameron at her home on the Isle of Wight in 1868 to sit for a portrait: he proclaimed it as one of his favourites.

This was a difficult topic to tackle, for religious orthodoxy lay against it. And even those prepared to accept that plant and animal species had evolved might refuse to believe that human beings could do so too, with their finer emotions such as love, altruism and religious sensibility emerging during descent from ape-like ancestors. Darwin knew that publishing these views would be controversial and for many years refrained, until the much less well-known naturalist and traveller Alfred Russel Wallace (1823–1913) came up with a parallel evolutionary theory, and time ran out.

On the Origin of Species (1859) laid out a clear stall. It was the summary of the work Darwin had been assimilating for two decades, writing to pigeon fanciers, market gardeners and dog breeders to collect examples from the experts on how heredity actually worked. Examining evidence first of all from the world of animal and plant breeders, Darwin asked the reader to agree that clearly domesticated animals or plants held within their bodies the potential to breed many different appearances not currently expressed. This mutability, Darwin argued, was used by human beings to create better agricultural strains. But in nature it suggested a way that a plant might accidentally express more of a trait than another, and then be better adapted to the environment in which it lived: it would then simply be more likely to survive and produce offspring.

Darwin's botanical interests continued in his *On the Various Contrivances by Which British and Foreign Orchids Are Fertilised by Insects, and On the Good Effects of Intercrossing* (1862) and then his 1875 book *Insectivorous Plants*, both of which were researched in the plant collections he himself amassed and cared for in his conservatory at Down House, in Kent, England. The illustrations in Darwin's orchid book demonstrated the most recent understanding of the way in which the form of flowers and their pollinators – for orchids, often moths – had co-evolved together. In earlier centuries, insects' visits had been seen as at best pointless and at worst damaging to seed and fruit; now, Darwin showed that the two were essential to each other.

ABOVE

***Angraecum sesquipedale*, *Curtis' Magazine*, (1859)**

Angraecum sesquipedale, pictured in *Curtis' Botanical Magazine* in 1859 – when the orchid was discovered, Darwin examined its long nectar tubes and correctly predicted it would need to have a pollinator with a 30cm proboscis.

Darwin also made predictions: when he was sent the Madagascar orchid *Angraecum sesquipedale* by James Bateman of Mexican orchid fame, Darwin noted it had long green nectar tubes hanging down behind the flower, surmising that there must exist a pollinator with a tongue that could reach to the bottom. 'Good heavens what insect can suck it,' he wrote. A few weeks later his prediction was more precise: 'A moth with probosces capable of extension to a length of between ten and eleven inches'. It would not be until 1907 that such a moth was identified in Madagascar. To acknowledge Darwin's achievement, it was named *Xanthopan morganii*, subspecies *praedicta*. It would take until 1992 for the moth to be recorded feeding on the flower in the wild.

The drama of the orchid itself is beautifully captured in hand-coloured nineteenth century lithographs, such as one from *Curtis' Botanical Magazine*. Curtis' had begun in 1787 (see also page 231) but by the mid 1800s had risen to be the preeminent horticultural periodical of record, and would have been an invaluable addition to the botanist's library of the time. To avoid the suggestion that the prints are merely aesthetic in aim, the artist has also included an impression of the whole plant's habit.

5 A place for women and working-class men?

The mid-nineteenth century saw the full professionalization of the sciences, ensuring that women and working-class people, limited in their access to education, had less access to positions of authority. Microscopes, laboratories, equipment and degrees became key to the practice of 'real science', and though women were still present as demonstrators, assistants, translators, illustrators and popularizers, it was extremely difficult for them to lead new scientific work.

There were a few exceptions. Anna Atkins (1799–1871) was given a scrupulously

TOP

Matilda Smith (1854–1926)

Matilda Smith, Joseph Hooker's cousin, who became one of the foremost botanical illustrators in Victorian London. She was the first to depict in depth the flora of New Zealand.

ABOVE

Joseph Dalton Hooker, photographed by Maull & Polyblank, (*c.*1855)

Shown here in his younger years, Hooker was in the process of publishing his botanical work from Tasmania and New Zealand, while acting as a sounding board for Darwin's new theories.

scientific education by her grieving father after her mother died postpartum. Atkins is considered the first person to have published a book with photographic images, which she made by printing distinctively blue cyanotype contact images of British seaweeds. She only made a few copies of her 1843 *Photographs of British Algae: Cyanotype Impressions* and it is therefore of extreme rarity value. In the 1850s she would go on to make several books about ferns by the same method.

But scientific posts were not always easily found, even for men. Darwin's science was always mediated through a network of correspondents, including Joseph Hooker (1817–1911), son of William (page 81). Hooker had no independent resources to travel as a gentleman-naturalist, and so had to join a naval expedition to the South Polar region in the late 1830s (page 65) as a salaried surgeon, giving him significantly less time to botanize. By mid-century Hooker had managed to secure himself a paid appointment, joining the Geological Survey looking at fossil plants in coal strata. It took wangling to secure funding from Kew Gardens and the Admiralty for a plant-collecting trip in 1847 to the Himalayas, although this would be the journey that would eventually make his reputation. Hooker's success in getting backing partly rode on the area's economic significance, given the strength of British commercial interests and ambitions in the region.

The books Hooker published were often subsidized, especially as regards their printing. It was not until the later half of the century, when Hooker had more liberty on a government stipend to publish, that he was able to cement the botanical reputation he arguably deserved. His *Flora of India* (seven volumes published from 1872) built on his Himalayan collecting, itself published in *Rhododendrons of Sikkim–Himalaya* (1849–51) and *Illustrations of Himalayan Plants* (1855), among others.

The plates for the latter book were prepared by Walter Hood Fitch (1817–92) who was *Curtis' Magazine*'s principal artist from the 1830s, but in the 1870s Joseph Hooker's daughter Harriet Hooker (1854–1945) also contributed, until 1880 when her cousin Matilda Smith (1854–1926) took over as principal. Smith painted for the magazine until 1923, as well as constantly producing images for Kew Gardens, and her contribution was significant enough that she was awarded the Silver Veitch Medal of the Royal Horticultural Society and in 1921 became only the second woman ever to be elected an associate by the Linnean Society. All of these women had grown up in an enthusiastically botanical context and were expert observers; however, they were seen as image-makers rather than scientists themselves.

The artists were only one part of the process of image-making. Walter Hood Fitch oversaw the plates for Hooker, from illustrations made by a Bengal civil servant, John Ferguson Cathcart (1802–51) using pictures made by local artists whose names we do not know. Cathcart had compiled over 1,000 images of Himalayan plants this way, but died in Lausanne before he was able to get back to London to oversee publication. The plates are exquisite, including botanical details as well as large images of the growing plants, and are considered to be some of the most beautiful ever produced.

A painting made by William Taylor in 1849 shows Hooker in the middle of his Himalayan journey, surrounded by indigenous people of the region, offering up their plants to the European expert. Yet this suggests a degree of separation from the process that is undermined by letters he wrote home: 'Botanizing during the march is difficult.

LEFT

The *Titan Arum*, or *Corpse Flower*, Matilda Smith, (1891)

The Titan Arum, or Corpse Flower, painted by Matilda Smith in 1891; the flower hails from Sumatra and was first described in 1878; it smells of rotting flesh to attract its pollinators.

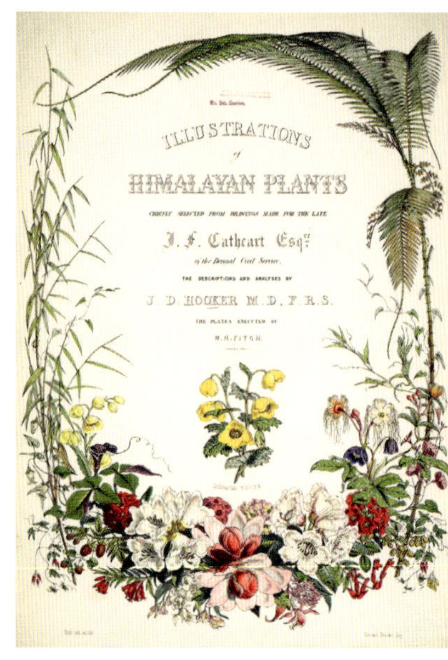

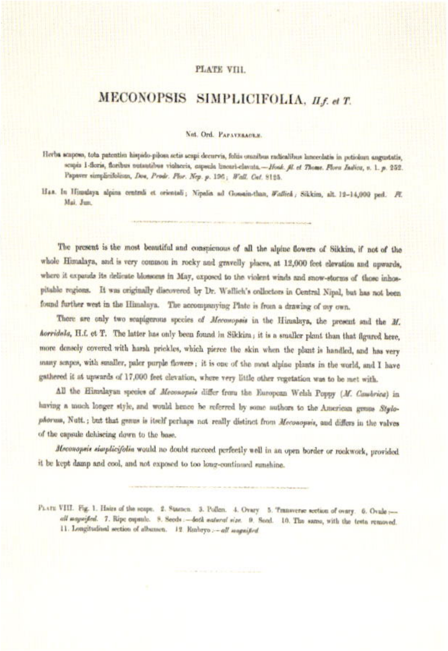

PLATE VIII.

MECONOPSIS SIMPLICIFOLIA, *H.f. et T.*

Nat. Ord. Papaveraceæ.

Herba scaposa, tota patentim hispido-pilosa setis scapi decurvis, foliis omnibus radicalibus lanceolatis in petiolum angustatis, scapis 1-floris, floribus nutantibus violaceis, capsula lineari-clavata.—*Hook. fil. et Thoms. Flora Indica*, v. 1. *p.* 252. Papaver simplicifolium, *Don, Prodr. Flor. Nep. p.* 196; *Wall. Cat.* 8125.

Hab. In Himalaya alpina centrali et orientali; Nipalia ad Gossain-than, *Wallich*; Sikkim, alt. 12–14,000 ped. *Fl.* Mai. Jun.

The present is the most beautiful and conspicuous of all the alpine flowers of Sikkim, if not of the whole Himalaya, and is very common in rocky and gravelly places, at 12,000 feet elevation and upwards, where it expands its delicate blossoms in May, exposed to the violent winds and snow-storms of those inhospitable regions. It was originally discovered by Dr. Wallich's collectors in Central Nipal, but has not been found further west in the Himalaya. The accompanying Plate is from a drawing of my own.

There are only two scapigerous species of *Meconopsis* in the Himalaya, the present and the *M. horridula*, H.f. et T. The latter has only been found in Sikkim; it is a smaller plant than that figured here, more densely covered with harsh prickles, which pierce the skin when the plant is handled, and has very many scapes, with smaller, paler purple flowers; it is one of the most alpine plants in the world, and I have gathered it at upwards of 17,000 feet elevation, where very little other vegetation was to be met with.

All the Himalayan species of *Meconopsis* differ from the European Welsh Poppy (*M. Cambrica*) in having a much longer style, and would hence be referred by some authors to the American genus *Stylophorum*, Nutt.; but that genus is itself perhaps not really distinct from *Meconopsis*, and differs in the valves of the capsule dehiscing down to the base.

Meconopsis simplicifolia would no doubt succeed perfectly well in an open border or rockwork, provided it be kept damp and cool, and not exposed to too long-continued sunshine.

Plate VIII. Fig. 1. Hairs of the scape. 2. Stamen. 3. Pollen. 4. Ovary 5. Transverse section of ovary. 6. Ovule:—*all magnified*. 7. Ripe capsule. 8. Seeds:—*both natural size*. 9. Seed. 10. The same, with the testa removed. 11. Longitudinal section of albumen. 12. Embryo:—*all magnified*

THIS PAGE

Illustrations of Himalayan Plants,
Joseph Hooker,
(1855)

The extended title page of the book explicitly mentions the artist John Cathcart, who died before publication, who had himself used the work of unknown Indian draughtsmen.

Sometimes the jungle is so dense that you have enough to do to keep hat & spectacles in company, or it is precipitous . . . certainly one often progresses spread-eagle fashion against the cliff, for some distance, & crosses narrow planks over profound Abysses, with no hand-hold whatever.'

As with earlier in the colonial period, indigenous people across the world were involved in the process of making botanical knowledge for outsiders, working as collectors, traders, brokers and experts. Of course, the power balance meant that often this relationship was exploitative, but locals held the knowledge on the ground. This was also true in Europe itself, where botanical 'outsiders' might exist very close to famous herbaria. One of the most intriguing areas of local expertise is those artisan botanists who were most at home in the pub.

Aspiring nineteenth-century minds might find interest in popular science lectures or publishing, but these would mostly have been produced by upper-middle-class interests. Recent scholarship has sought to assert the importance of expert botanists from artisan backgrounds who met often in pubs, partly to find a common room in which a box might be locked to safeguard the society's books and herbarium. 'We instruct one another by continually meeting together; so that the knowledge of one becomes the knowledge of all', said John Horsefield (1792–1854), the president of the Prestwich Botanical Society, based near the growing northern city of Manchester.

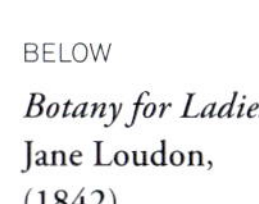

BELOW

Botany for Ladies,
Jane Loudon,
(1842)

Botany for Ladies, an introduction in English for those who hadn't learnt Latin at school, was published by Mrs Jane Loudon in 1842; she declared it a 'natural system', following de Candolle, in opposition to the Linnean classification which was falling out of fashion.

These keen individuals collected new species, identified those common to the locality and shared in a common practice of botanizing. One practice was insisted upon: the naming of species in Latin by Linnaean nomenclature. Horsefield's colleague Leo Grindon (1818–1904) made a herbarium substantial enough to form the basis of Manchester Museum when it was established in 1860. Grindon's *Manchester Flora* of 1859 is the kind of book that would have been available to such botanizers, spending time in the introduction explaining in simple English terms the necessary botanical concepts, with clear communicative illustrations of differences. Practising botany and gaining scientific expertise may have been a way for these highly skilled manual workers to reclaim some sense of pride and skill as their professional lives were threatened by the coming of mechanization.

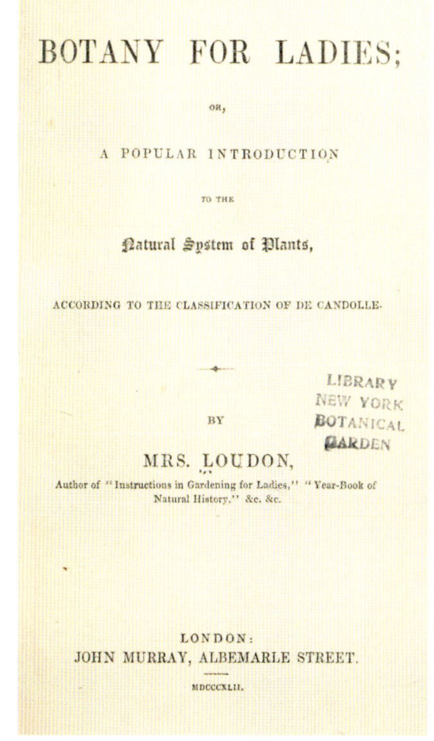

BOTANY FOR LADIES;

OR,

A POPULAR INTRODUCTION

TO THE

Natural System of Plants,

ACCORDING TO THE CLASSIFICATION OF DE CANDOLLE.

BY

MRS. LOUDON,

Author of "Instructions in Gardening for Ladies," "Year-Book of Natural History," &c. &c.

LONDON:
JOHN MURRAY, ALBEMARLE STREET.

MDCCCXLII.

The issue of how to make botany available to a wider sphere of people was a thorny one. Jane Loudon (1807–58) wrote *Botany for Ladies* in 1842, which bemoaned the difficulty of many botanical texts. Such books she said were 'sealed books' to her, and even John Lindley's *Ladies' Botany* (1834) 'did not tell me half I wanted to know, though it contained a great deal I could not understand'. Loudon described how the books seemed to have been written with the assumption that a teacher would be on hand to explain them. The modern audience, for her, demanded something they really could use to teach themselves. The debate about the 'right' language to use to describe parts of plants persisted – university-level floras today tend to eschew images in favour of descriptive paragraphs.

BELOW

***Himalayan Journals*, Joseph Hooker, (1854)**

View of 'Kanchanjunga from Hodgson's Bungalow', from Joseph Hooker's Himalayan Journals, published 1854.

THIS PAGE

Cyanotype images, Anna Atkins

These four images show the work of Anna Atkins, who pioneered the cyanotype, which gives these distinctive blue silhouetted shapes. Top left *Spiraea aruncus*, from the Tyrol, top right, *Eryngium*, a member of the sea hollies. Both images are taken from a a unique album in the Getty Museum. Left and right lower, two of her title pages; *Photographs of British Algae* is considered to be the first book illustrated with photographs.

ABOVE

Jane Loudon
(1807–1858)

Jane Loudon, botanical writer and also early pioneer of science fiction, she met her husband, the older botanist John Loudon, when he wrote a favourable review of her futuristic novel, *The Mummy*.

RIGHT

Ladies Flower Garden Annuals,
Jane Loudon
(1843)

This delicate plate shows an *Auricula*, a favourite amongst Victorian flower enthusiasts.

OPPOSITE

Ladies Flower Garden of Ornamental Perennials,
Jane Loudon
(1849)

A hand-coloured chromolithograph, a new printing technique at the time, showing varieties of *Dahlia superflua*, the variable dahlia, the Sir Robert Peel, Harris' Inimitable, and Levic's Incomparable.

Pl 46
1 Dahlia superflua (the origin of all the garden varieties) _2 the Sir Robert Peel _3 Harris' Inimitable
4 Lewis' Incomparable

THIS PAGE AND OPPOSITE

Ladies Flower Garden of Ornamental Perennials, **Jane Loudon** (1849)

Left, *Anemone coronaria*, and *Amaryllis belladonna*, right.

3
1
2
1. Belladonna purpureus. 2. Belladonna blanda. 3. Belladonna purpureus pallida

RIGHT

Coloured lithograph British Ferns, unknown artist, (*c.*1885)

Increasing enthusiasm for fern-collecting became a 'craze' in mid-Victorian times, leading to a market for anything fern-related.

OPPOSITE

***The Instructive Picturebook, or Lessons from the Vegetable World,* Charlotte Mary Yonge, (1858)**

Lichen, fungus, alga, moss, fern, grass and palm, from this educational work for younger Victorian readers.

b
a
e
c
d
f
g
a Lichen.
b Fungus.
c Alga.
d Moss.
e Fern.
f Grass.
g Palm.

6 Bentham and Hooker and the growth of botanizing

With every new generation of plant science, practitioners have introduced new methods of organizing the plant world, believing that their innovations solve the problems of those that came before. Joseph Hooker's career saw him spend years at Kew Gardens gatekeeping (and mostly rejecting) the arrivals of newly-named 'species' from enthusiastic botanizers around the globe. In his maturity he began work on a full system of classification for plants with his older colleague George Bentham (1800–84), based on the idea that natural similarities should lie at the heart of taxonomy.

Bentham was a keen botanist in his youth but had to work elsewhere until his uncle died, leaving him financially independent. As soon as Bentham was freed to travel he visited every European herbarium he could, compiling and publishing on plant species. From 1857 he was employed on a government mission to create floras of the many British colonies and possessions globally, beginning with the *Flora Hongkongensis* of 1861, and later the *Flora Australiensis* (*Flora of Australia*, 1863–1878), the first botanical survey of a whole land area to be completed.

However, it was their shared endeavour the *Genera plantarum* (1862–3), which engraved the book's nickname – 'Bentham and Hooker' – in botanical memory. The book covered the seed plants only (not mosses or ferns). As if to reaffirm the idea of botany as a serious science, bounded by educational standards, it appeared in Latin and there were no illustrations, instead organized by careful description of some 97,000 species into about 200 family groupings. The classification was considered substantial enough to be used until late in the twentieth century (for example, Kew Gardens' order beds were arranged by its findings).

BELOW

George Bentham, copy of original by Lowes Dickinson, **attributed to Emily Merrick, (19th century)**

Bentham was a conscientious and tireless worker, who lived his life by the principle that all he wrote should have been observed at firsthand.

Botany was becoming a serious subject in universities with its own courses, now no longer under the medical faculty. The nineteenth century had seen increasing numbers of lecturers using illustrations during their work, such as John Stevens Henslow. Sometimes this was essential, as nature did not always cooperate in displaying species in tandem with the lecture calendar. Daniel McAlpine (1849–1932) was a Scottish botanist who emigrated to Australia and became the empire's first professional plant pathologist. His *Botanical Atlas* of 1883 showed a university-level student how to see the plant itself, labelling each part with pedagogical care. Students were encouraged to draw the plants themselves, with teachers believing that in making a drawing they would see more clearly what was and was not there.

In addition, amateur botany was becoming increasingly popular, and attention to the living plant was encouraged here too. Phoebe Lankester (1825–1900) was an important champion of this method, beginning with her *A Plain and Easy Account of the British Ferns* (1860). Her books did not contain very notable illustrations, but she believed in drawing a plant as a way of learning about it. 'I am glad to write again of a sea-shore plant' she said of sea holly, 'for I associate with them quiet bright holiday hours'. Her work keyed into another central enthusiasm of the mid-Victorians – an enthusiasm for the collecting of the natural world.

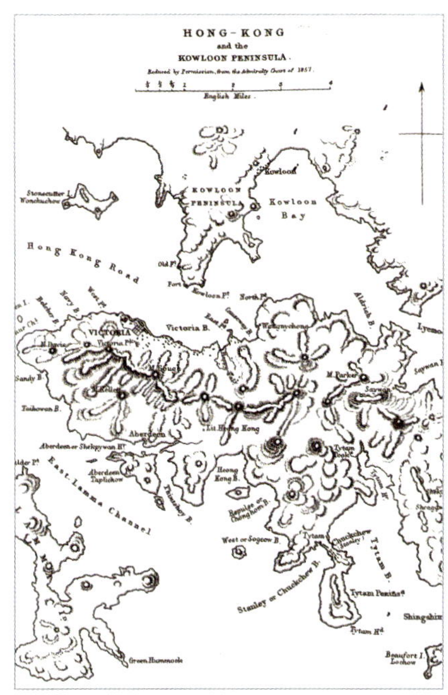

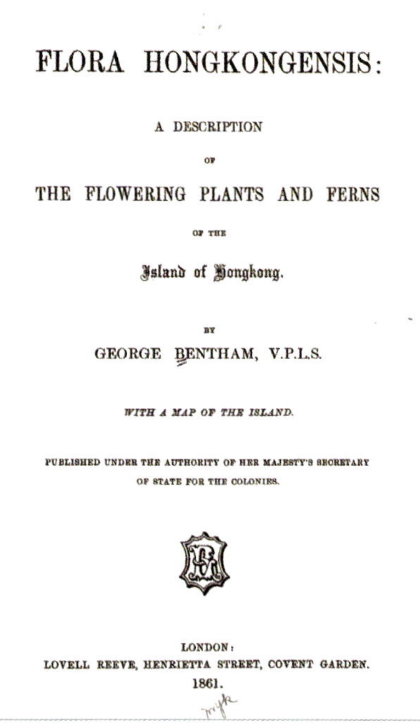
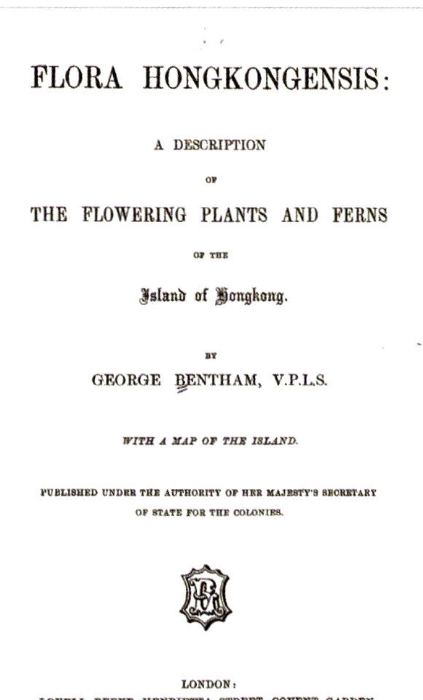

FLORA HONGKONGENSIS:

A DESCRIPTION

OF

THE FLOWERING PLANTS AND FERNS

OF THE

Island of Hongkong.

BY

GEORGE BENTHAM, V.P.L.S.

WITH A MAP OF THE ISLAND.

PUBLISHED UNDER THE AUTHORITY OF HER MAJESTY'S SECRETARY OF STATE FOR THE COLONIES.

LONDON:
LOVELL REEVE, HENRIETTA STREET, COVENT GARDEN.
1861.

LEFT

Flora Hongkongensis, Genera Plantarum, and *Flora Australiensis,* George Bentham (1861)

Bentham was instrumental in creating a series of floras that spanned the British Empire.

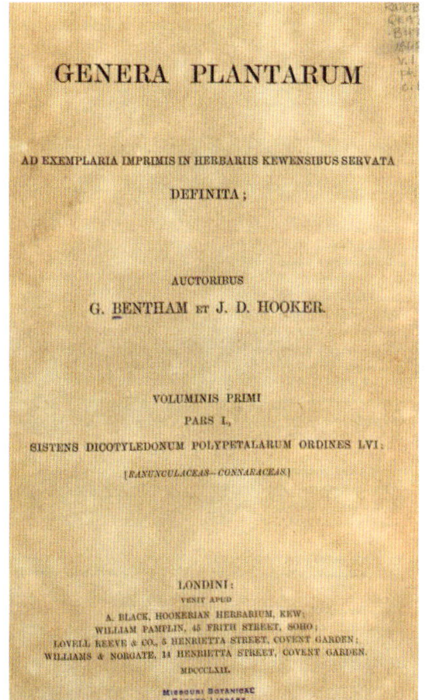

GENERA PLANTARUM

AD EXEMPLARIA IMPRIMIS IN HERBARIIS KEWENSIBUS SERVATA DEFINITA;

AUCTORIBUS
G. BENTHAM ET J. D. HOOKER.

VOLUMINIS PRIMI
PARS I,
SISTENS DICOTYLEDONUM POLYPETALARUM ORDINES LVI:
[*RANUNCULACEAS—CONNARACEAS.*]

LONDINI:
VENIT APUD
A. BLACK, HOOKERIAN HERBARIUM, KEW;
WILLIAM PAMPLIN, 45 FRITH STREET, SOHO;
LOVELL REEVE & CO., 5 HENRIETTA STREET, COVENT GARDEN;
WILLIAMS & NORGATE, 14 HENRIETTA STREET, COVENT GARDEN.
MDCCCLXII.

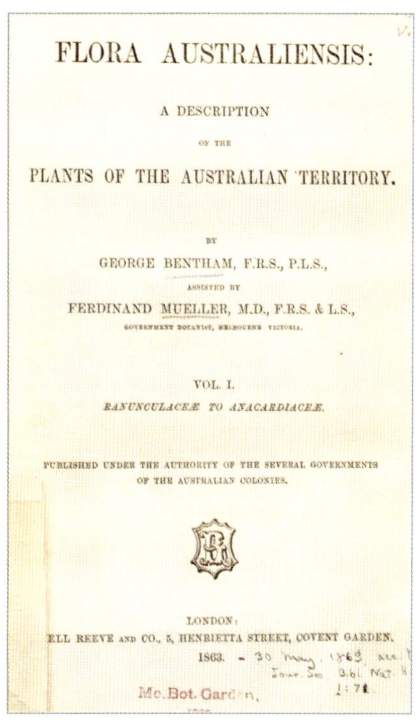

FLORA AUSTRALIENSIS:

A DESCRIPTION

OF THE

PLANTS OF THE AUSTRALIAN TERRITORY.

BY

GEORGE BENTHAM, F.R.S., P.L.S.,

ASSISTED BY

FERDINAND MUELLER, M.D., F.R.S. & L.S.,

GOVERNMENT BOTANIST, MELBOURNE, VICTORIA.

VOL. I.

RANUNCULACEÆ TO ANACARDIACEÆ.

PUBLISHED UNDER THE AUTHORITY OF THE SEVERAL GOVERNMENTS OF THE AUSTRALIAN COLONIES.

LONDON:
ELL REEVE AND CO., 5, HENRIETTA STREET, COVENT GARDEN.
1863.

BELOW

Genera Plantarum (The Genera of the Plants), George Bentham and Joseph Hooker, (1902)

Bentham will be most remembered for *Genera Plantarum (The Genera of the Plants)* written with Joseph Hooker, an attempt to classify all known plants of the globe into a family pattern.

This was most clearly seen in the fern craze, described by scholar of natural history, David Allen, as 'the most destructive natural history fashion of all'. The start of the nineteenth century saw a fashion for tourism and for experiences that suggested leaving a conventionally English landscape – climbing mountains, walking in forests. Fern-filled gullies were created throughout the British Isles, such as that at Belfast Botanic Garden, from 1889, or the private fernery at Ascog in Bute, from 1875. Many of the plants in such ferneries were collected from the wild.

Books probably helped this passion along significantly, for example, *An Analysis of the British Ferns and their Allies* (1837) by George William Francis (1800–65), who later emigrated to become the first director of Adelaide Botanic Garden, and *A History of British Ferns* (1840), by Edward Newman (1801–76), editor of English hunting magazine *The Field*. However, the hobby remained relatively specialist until the 1850s when popular publishing exploded with guides to fern collecting, and newspapers began to note that the countryside was being churned up in search of species.

In the wake of the new taste for the gothic, ferns appeared on fabric and furniture, Christening presents and gravestones, and even on the new design for the surface of custard cream biscuits. Nona Bellairs (1824–97) who wrote botanical guidebooks combining travel accounts of customary monuments, such as the English Church in Nice, with observations of ferns, said quite passionately, 'We must have "Fern laws", and preserve them like game.' Other English-speaking parts of the world also fell prey to fern fever, to an extent in America but very noticeably in Australia, where railway stations and other public buildings might be festooned in collected wild fern specimens. Even William Hooker himself published a fern book, *A Century of Ferns* (1854), with beautiful hand-coloured images.

7 Mendel, heredity and the beginning of genetics

Austrian-Czech Gregor Mendel (1822–84) was an important exception to the movement towards laboratory-based science, which may explain why his work was little-known during his lifetime. He lived his adult life as a monk in St Thomas' Abbey, Brno (now in the Czech Republic), where he took charge of a 2-hectare (5-acre) experimental garden, cultivating plants and observing how they transmitted characters from one generation to the next. He chose the traits he would observe, eventually settling on peas as a good experimental subject, and from 1856–63 he grew and recorded results for 28,000 plants.

The patterns he recorded allowed him to observe some general rules for how heredity worked, which he published at the local natural history society meeting proceedings. Mendel observed that a second generation has a 3:1 ratio in expressing a particular character, such as flower colour. He did not know why this happened, but he saw it was a pattern in the natural world. He also realized that one plant might inherit colour from a parent, but colour of the seeds from the other. This is the law of independent assortment. Mendel looked at the way grandchildren can inherit from their grandparents and theorized that the parental generation must carry coded information not expressed in themselves, but reappearing in their offspring.

Although Mendel's published work set out clear rules for the process of inherited characteristics, his audience did not understand the significance of what they were

GREGOR MENDEL

Abbot of Brünn

Born 1822. Died 1884.

From a photograph kindly supplied by the Very Rev. Dr Janeischek, the present Abbot.

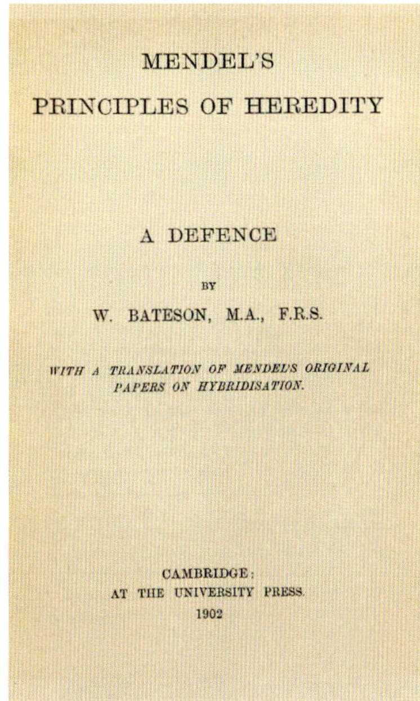

MENDEL'S

PRINCIPLES OF HEREDITY

A DEFENCE

BY

W. BATESON, M.A., F.R.S.

WITH A TRANSLATION OF MENDEL'S ORIGINAL PAPERS ON HYBRIDISATION.

CAMBRIDGE:
AT THE UNIVERSITY PRESS.
1902

320 APPENDIX

observation completely or cannot be detected with certainty. This circumstance is of great importance in the determination and classification of the forms under which the offspring of the hybrids appear. Henceforth in this paper those characters which are transmitted entire, or almost unchanged in the hybridisation, and therefore in themselves constitute the characters of the hybrid, are termed the *dominant*, and those which become latent in the process *recessive*. The expression "recessive" has been chosen because the characters thereby designated withdraw or entirely disappear in the hybrids, but nevertheless reappear unchanged in their progeny, as will be demonstrated later on.

It was furthermore shown by the whole of the experiments that it is perfectly immaterial whether the dominant character belongs to the seed-bearer or to the pollen-parent; the form of the hybrid remains identical in both cases. This interesting fact was also emphasised by Gärtner, with the remark that even the most practised expert is not in a position to determine in a hybrid which of the two parental species was the seed or the pollen plant.[1]

Of the differentiating characters which were used in the experiments the following are dominant:

1. The round or roundish form of the seed with or without shallow depressions.
2. The yellow colouring of the seed albumen [cotyledons].
3. The grey, grey-brown, or leather-brown colour of the seed-coat, in association with violet-red blossoms and reddish spots in the leaf axils.
4. The simply inflated form of the pod.
5. The green colouring of the unripe pod in association with the same colour in the stems, the leaf-veins and the calyx.
6. The distribution of the flowers along the stem.
7. The greater length of stem.

With regard to this last character it must be stated that the longer of the two parental stems is usually exceeded by the hybrid, a fact which is possibly only attributable to the greater luxuriance which appears in all parts of plants when stems of very different length are crossed. Thus, for instance, in repeated experiments, stems of 1 ft. and 6 ft. in length yielded without exception hybrids which varied in length between 6 ft. and 7½ ft.

[1] [Gärtner, p. 223.]

reading. It was not until the early twentieth century that those beginning to build the early science of genetics finally found Mendel's work and saw him as a rich predecessor. Hugo de Vries (1848–1935) in the Netherlands, suggested in his *Die Mutationstheorie* (*The Mutation Theory*, 1901) that perhaps the way an organism appears is determined by particles, each containing separate instructions for a particular physical expression, such as flower colour.

De Vries independently went through the same kind of plant experiments as Mendel, having selected *Oenothera* (evening primrose) as his subject, and observed the same 3:1 ratio of appearances in the second generation. The recessive gene, though carried by other individuals, will only be expressed in those that carry two copies. De Vries published a full account of his work in 1901–03, which came with twenty-two lithograph illustrations; not only of the flowers concerned, but also of his experimental beds.

The discoveries of Mendel and the subsequent generation were enthusiastically taken up by plant breeders, who believed this new science might be of great use. Luther Burbank (1849–1926) was a pioneer of horticultural and agricultural science, establishing one of the largest scientific research farms in the USA. Early in his career he bred a blight-resistant potato called the russet Burbank, which remains the most grown in America, and from which most McDonald's fries are made. He was a true Californian, embracing yogic initiation from Paramahansa Yogananda, but his life's work was rewarded in the form of the Plant Patent Act, 1930, which allowed companies to protect the intellectual property of new breeding efforts.

ABOVE

Gregor Mendel, from the frontispiece to English biologist William Bateson's English introduction to his book *Mendel's Principles of Heredity: A Defence*, (1902)

This central text for the understanding and development of genetics was only published in the early twentieth century, once Mendel's research had been rediscovered.

8 The botanist gardener and the new interest in the wild

In the later years of the nineteenth century, English gardening began to take a turn away from exotic introductions and towards more old-fashioned cottage plants. William Robinson (1838–1935) was born in Ireland but emigrated to London to work at the botanical gardens of Regent's Park, specializing in native wildflowers. In 1866 he became a fellow of the Linnean Society, his application sponsored by Charles Darwin and James Veitch, the nursery owner. Robinson started his own journal, *The Garden*, publishing articles which showed a post-Darwinian understanding of the relationships between plants and their wider environment, including twenty-six by Alfred Russel Wallace.

Robinson wrote *The Wild Garden* (1870), and his beliefs were expressed at his home, Gravetye Manor in Sussex, where large drifts of bulbs such as narcissi were planted under the woodland canopy, so that the garden spread outwards, its edges not immediately apparent. Even his most formal beds around the house were planted according to more ecological principles, filled to the brim so that there was no soil showing. In an article for *The Garden* magazine in 1872, called 'What is a Wild Garden?', a black and white illustration shows a scene with little evidence of having been gardened at all. The illustrations to his work show a fondness for plants shown in their growing context, rather than singly.

Robinson was an enthusiastic collector of alpines, and described his journeys to collect plants in Europe and the Rocky Mountains in his *Alpine Flowers for Gardens: Rock, Wall, Marsh Plants, and Mountain Shrubs* (1870). The first edition of the book came with a spine illustration of small alpine plants, plus rustic lettering constructed from drawn twigs. His alpine enthusiasm suggests how his interest in wild species went beyond the British natives to others that fitted his aesthetic.

Robinson had a strong relationship with his gardener colleague Gertrude Jekyll (1843–1932). Jekyll was preoccupied with the evolving landscape in England, particularly her home county of Surrey, where she saw life rapidly changing in the face of modernity. In *Wood and Garden* (1904) even the way in which Jekyll organized her text, less a catalogue and more a gentle walk to see what might turn up, month by month, suggested a different way of seeing plants growing in the world: as a whole community, similarly adapted to its environment, rather than as single analyzable specimens. We can see here the hints of a more holistic view of nature.

In Denmark, Eugen Warming (1841–1924) exemplified a similar shift, this time in university botany. Having held positions at Stockholm and Copenhagen universities, in 1895 Warming published a book of his lectures called *Plantesamfund* (*Plant Communities*), which showed how plants adapted to different environmental conditions and habitats, often by similar adaptive mechanisms. Warming's lectures were soon translated into English, at which point the word 'oecology' was added to the title, the first time the word had been used.

Warming's work was inspirational to a generation of young scientists including Arthur Tansley (1871–1955) and Frederic Clements (1874–1945). For the first time, botanists were studying plants within their natural habitats, in the associative communities they tend to form. Such young ecologists aimed to create a science that tracked how energy

TOP

Gertrude Jekyll in the garden at the Deanery, Sonning, photographer unknown but probably Maxwell-Lyte (*c.*1905)

Jekyll designed the garden at The Deanery, a house by her long-time collaborator, Edwin Lutyens; it expressed her interest in plants looking more naturally situated in their surroundings.

OPPOSITE BOTTOM

William Robinson, garden designer, in a photograph from around 1930

Robinson was greatly interested in the way plants grew in the wild, and especially how they adapted to different habitats.

BELOW

***Hibiscus archeri*, a coloured plate (1899)**

Robinson's career embraced garden-making, authorship and magazine editorship; vivid coloured images such as these helped to make his work a commercial success.

ABOVE

Arthur Tansley in 1893, while still studying at Cambridge

Tansley pioneered the idea of an 'ecosystem' and was passionate about creating organizations to safeguard the British countryside.

moved through natural habitats, harnessed from the sun by plants and then grazed by animals who would then themselves be hunted. The view was of a system, rather than of individual species.

Tansley became Professor of Botany at the University of Oxford, and helped to found the British Ecological Society. Tansley's work made him committed to the idea of conservation and after the Second World War he worked to form the Nature Conservancy, providing legal protection for significant botanical reserves under the 1949 National Parks Act. Clements, in America, became interested in how a denuded habitat might progress through stages to what he christened a 'climax community', a stable set of plant species. And it was Aldo Leopold (1887–1948) who took the argument for conservation a stage further, arguing that all species had a philosophically equal right to exist. He argued against road building into the wilderness, against the utilitarian use of land resources (for example, in hydroelectric dams) and for the value of apex predators such as wolves.

RIGHT

***Oecology of Plants*, by Eugen Warming, first published in Danish in 1895, in its first English version, (1909)**

This book was the first use of the word 'ecology', albeit with its original spelling, and influenced a whole generation of botanists.

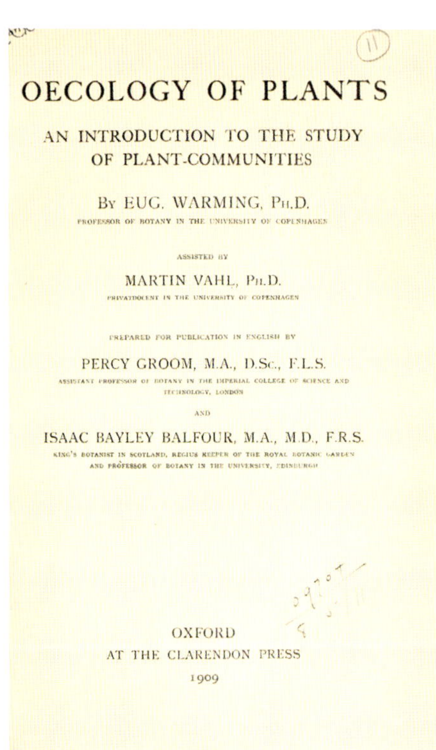

OECOLOGY OF PLANTS

AN INTRODUCTION TO THE STUDY OF PLANT-COMMUNITIES

BY EUG. WARMING, PH.D.

PROFESSOR OF BOTANY IN THE UNIVERSITY OF COPENHAGEN

ASSISTED BY

MARTIN VAHL, PH.D.

PRIVATDOCENT IN THE UNIVERSITY OF COPENHAGEN

PREPARED FOR PUBLICATION IN ENGLISH BY

PERCY GROOM, M.A., D.SC., F.L.S.

ASSISTANT PROFESSOR OF BOTANY IN THE IMPERIAL COLLEGE OF SCIENCE AND TECHNOLOGY, LONDON

AND

ISAAC BAYLEY BALFOUR, M.A., M.D., F.R.S.

KING'S BOTANIST IN SCOTLAND, REGIUS KEEPER OF THE ROYAL BOTANIC GARDEN AND PROFESSOR OF BOTANY IN THE UNIVERSITY, EDINBURGH

OXFORD

AT THE CLARENDON PRESS

1909

GENUS 7. FERN FAMILY. 21

9. Dryopteris Filix-más (L.) Schott. Male Fern. Fig. 45.

Polypodium Filix-mas L. Sp. Pl. 1090. 1753.
Aspidium Filix-mas Sw. Schrad. Journ. Bot. 1800²: 38. 1801.
Dryopteris Filix-mas Schott, Gen. Fil. 1834.

Rootstock stout, woody, ascending or erect, chaffy. Leaves up to 4° high, in an erect crown; stipes 4'–10' long, densely chaffy below; blades nearly evergreen, 1°–3° long, 6'–11' broad, broadly oblong-lanceolate, acuminate, narrowed at the base, nearly or quite 2-pinnate; pinnae narrowly deltoid-lanceolate to oblong-lanceolate, acuminate; segments adnate, oblong, obtuse and biserrate, or partially adnate, ovate-oblong, acutish and deeply incised; sori numerous, large, nearer the midvein than the margin; indusia orbicular-reniform, glabrous.

In rocky woods, Newfoundland and Labrador to Alaska, south to Vermont, northern Michigan, South Dakota, Arizona and California. Aug. Also in Greenland. Numerous related forms of wide distribution are referred to this species; the type is European. The rootstock of this and the preceding species furnish the drug Filix-mas used as a vermifuge. Basket-fern. Male shield-fern. Shield-roots. Bear's-paw-roots. Sweet or knotty brake.

10. Dryopteris spinulòsa (Muell.) Kuntze. Spinulose Shield-fern. Fig. 46.

Polypodium spinulosum Muell. Fl. Fridr. 113. *f.* 2. 1767.
Aspidium spinulosum Sw. Schrad. Journ. Bot. 1800²: 38. 1801.
Dryopteris spinulosa Kuntze, Rev. Gen. Pl. 2: 813. 1891.

Rootstock stout, creeping, chaffy. Leaves in an incomplete crown, the taller erect, the others spreading, stipes 4'–14' long, with pale brownish scales; blades ½°–1½° long, 3½'–9' broad, ovate-lanceolate to oblong, acuminate, deeply 2-pinnatifid; pinnae usually oblique, pinnately divided, the lower ones unequally deltoid, those above lanceolate from a broad base, acuminate; pinnules flat, oblong to lanceolate, acute, decurrent, pinnately cut, segments incised, teeth mucronate, falcate, appressed; sori submarginal, terminal on veinlets; indusia without glands.

In rich low woods, Labrador to Selkirk and Idaho, to Virginia and Kentucky. Also in Europe. Called also Narrow Prickly-toothed Fern.

11. Dryopteris dilatàta (Hoffm.) Gray. Spreading Shield-fern. Fig. 47.

Polypodium dilatatum Hoffm. Deutschl. Fl. 2: 7. 1795.
Aspidium spinulosum var. *dilatatum* Hook. Brit. Fl. 444. 1830.
Dryopteris spinulosa var. *dilatata* Underw. Nat. Ferns, ed. 4, 116. 1893.

Rootstock creeping, or ascending. Leaves equal, spreading, in a complete crown; stipes ½°–1½° long, with dark brownish often darker-centered scales; blades ⅔°–2½° long, 4'–16' broad, triangular to ovate or broadly oblong, acuminate, 3-pinnatifid; pinnae variable, the lower ones broadly and unequally ovate or triangular, those above lanceolate to oblong, acute or acuminate, the lowermost at least pinnately divided; pinnules convex, oblong to lanceolate, acute, the largest not decurrent, pinnately divided, segments pinnately lobed, teeth mucronate, straight or falcate, usually not appressed; sori mostly subterminal; indusia glabrous, or with a few glands.

A high mountain species of rocky woods, Newfoundland to Alaska, California, Idaho, Tennessee and North Carolina, Greenland. Also in Eurasia, Japan and the Madeira Islands. Broad Prickly-toothed Wood-fern.

FAR LEFT

Aldo Leopold, Professor at the University of Wisconsin

Leopold spent his life working to preserve American wilderness, as a government employee, an academic, and in his private work as a writer.

LEFT

***A Sand Country Almanac*, by Aldo Leopold, first published 1949**

Leopold finished his manuscript not long before his death, and did not live to see it sell over two million copies in more than a dozen languages.

BELOW

Binoculars and a travel case that belonged to Leopold, now on display in his former cabin

Leopold bought wild land in Wisconsin, an experience which he wrote about in his book; the land is now a Memorial Reserve.

ABOVE

John Muir, campaigner for the establishment of US National Parks

Born in Scotland in 1838, he stands amongst the tall trees now so associated with his name. California, home of Muir Woods National Park, celebrates John Muir Day on 21 April each year.

9 Conservationism begins and the threat of war

John Muir (1838–1914) was one of the founding spirits of modern conservation. His view of the natural environment, and of the experience of being in it, has shaped a view of the world that sees us not as separate entities but one whole. His first triumph was the preservation of Yosemite as the first US National Park, using the press to call on Congress for protective legislation.

Muir began botanizing in earnest after having moved to Ontario in 1863, to avoid the draft for the American Civil War. It was a dream to finally find work in 1878 as an official artist on the government survey of the 39th parallel. Muir lived in the area that would become Yosemite, working as a shepherd and building himself a hut, and also recording the giant sequoia trees, an experience that is described in his book *My First Summer in the Sierra* (1911).

Muir argued that a national park should be a place of spiritual retreat, without use value, while his opponents argued it should be made use of for human benefit. He fundamentally believed in the idea of wilderness as opposed to humanity, and that wildness is inherently superior, simply by being wild. The debate continues even today, but his influence shaped the dialogue. Mongolia lays claim to the creation of the first ever national park, at Bogd Khan Uul, a century earlier in 1783, but Muir sparked Western feeling about the subject.

This also marks a foundational moment for popular conservation movements. In England at almost exactly the same time, in 1895, the National Trust was founded. An important figure was Beatrix Potter (1866–1943), who, before becoming a children's book author, had become deeply interested in fungi, to the point of submitting a paper to the Linnean Society about it. Her work illustrated *Wayside and Woodland Fungi* (1967) by W.P.K. Finlay (1904–1985). On her death, she donated her farmland, 160 hectares (4,000 acres) carefully accumulated by a series of parcel purchases, to the National Trust.

As the nineteenth century waned, European war lay on the horizon once more. British concerns had been raised during the Boer War in South Africa (1899–1902) when many young men failed to pass the most basic tests of fitness to serve. At the same time, the growth of the global economy threatened national fruit crops, as apples were shipped from France and later even America on huge refrigerated ships. The fruit orchards that had surrounded London were disappearing in favour of suburban house-building, and even the traditional weights and measures used at Covent Garden market were no longer of use to the new buyers and sellers, often being too individually attached to particular crops.

A growing movement sought to record and preserve the world of British fruit, fearing a mortal injury to the British national body. George Brookshaw's *Pomona Britannica* (1812) had been a prescient precursor of this (a pomona is a book devoted to recording fruit). Then in 1851, Dr Robert Hogg (1818–97) published his *British Pomology*, expressing some of the same concerns, and by the 1870s Hogg and his colleague Dr Henry Graves Bull (1818–85), a prison doctor and fungi expert, had organized *The Herefordshire Pomona*, an effort to record all of the lovely local apple species of that county. The *Pomona* was published in parts, but the collected edition, bound in green Morocco leather, is a true collector's item, full of

ABOVE

Beatrix Potter outside her Lake District home in 1913

With the money made from her world-famous children's books, Potter bought farmland in the Lake District, which became her family home.

LEFT

***Hygrophorus puniceus* (above) and *Himeola auricula* (left).**

In her twenties Potter became fascinated by mycology, the study of fungi, and lent her delicate watercolour style to many paintings.

THIS PAGE

Lepiota friesii **(right) and *Strobilomyces strobilaceus*** **(below right)**

Potter was unable to make inroads in the overwhelmingly masculine life of Victorian scientific London, and her work on fungi was not acknowledged, despite the attentive nature of her work.

LEFT

Pomona Britannica, George Brookshaw, (1817)

The blossom, leaves and fruit of Nectarine (*Prunus persica*), shown here in tempting hand-stippled engravings.

RIGHT AND OPPOSITE

Pomona Britannica,
George Brookshaw,
London
(1817)

Plate showing three Cherry (*Prunus avium*) varieties: George II's Cherry, Graffien or Biggarou, and the gloriously-named Harrison's Heart. Raisin de Calmes grape (opposite), *Vitis vinifera*. Growing grapes and other exotic fruits was increasingly a high-status occupation for the wealthy of the mid-nineteenth century.

glorious chromolithographic illustrations of the fruit, made from paintings by Alice Blanche Ellis (fl. 1876–1916) and Edith Elizabeth Bull (dates unknown). Even a copy in poor condition can fetch over £10,000. Dr Bull's drawings of fungi are held at Kew Gardens.

National floras became more important in an era of growing nationalist feeling. In France, Gaston Bonnier (1853–1922) made life easier for the amateur, even one who wanted to range into neighbouring Belgium, Germany and Switzerland – his *Flore complète illustrée de France, Suisse et Belgique* (*Complete Illustrated Flora of France, Switzerland and Belgium*) of 1911–34, written with Robert Douin (1892–1965), brought this war-torn region into one botanical unity, advertising his work as '*sans mots techniques*' – 'described without technical words'. The illustrated pages hold lovely and lively paintings of several flowering plants on each double spread (much revised and updated, this remains one of the standard floras of the region even today, still bearing Bonnier's name, most recently updated in 1985).

RIGHT

***The Herefordshire Pomona*, illustrated by Alice Evans and Edith Bull, (1878–84)**

Evans and Bull together produced over 400 immaculate watercolours for reproduction by the specialist printers, ironically, given the book's concerns to record rare British varieties, located in Belgium.

Frederick Warne (1825–1901), the successful London publisher, began creating the *Observer* book series just before the start of the Second World War, in 1937; the first title in the series was birds, but the second was *British Wild Flowers*, and mint copies of this now fetch several hundred pounds. Amazingly, even during wartime, while *Observer* were preparing a rushed book on aeroplanes to help people spot enemy aircraft, the publishing house also managed to turn out the far more gentle and soothing-sounding *British Grasses, Sedges and Rushes* (1942).

These images encouraged a new generation of observers to think of themselves as botanically minded, searching out species and trying to identify them. But it also motivated them to think of botany as a worthwhile leisure activity, along the same lines as their Victorian forebears, though with far less 'moral' judgment attached. For these botanizers, the value was to be found in being in the open air, exploring new areas of their own country, experiencing 'the outdoors' and a consequent sense of freedom. Bicycles and later cars enabled such enthusiasts to roam further.

10 The laboratory, the world and the future

Meanwhile in the laboratory, an increasing amount of botanical work studied plants at the cellular level. The German scientist Richard Willstätter (1872–1942) had theorized that the chlorophyll molecule was absorbing light energy, which would then go through a series of reactions to become sugar (he later won the Nobel Prize for this work). However, it was nuclear physicists such as the American Ernest Lawrence (1901–58), inventor of the cyclotron, who suggested the possibility of using carbon-14, a relatively long-lived radioactive isotope of carbon, as a tracer. In the 1940s, Melvin Calvin (1911–97) and

Andrew Benson (1917–2015) used the distinctive carbon to track the processes by which plant cells take in light energy, before catalyzing reactions with carbon dioxide, to store the energy in the form of sugars. Calvin won the 1961 Nobel Prize for this work, with a crucial insight that came to him at a quiet moment, 'waiting in my car while my wife was on an errand'.

Elsewhere in botany, the wider world was wrangling with the West's historical botanical dominance. Japanese botanists, for example, found frustration when the defining type specimens of native species were held in Europe and were described in just a few lines, unhelpful for Japanese queries. The implication seemed that any botanists with inquiries could easily access the big European herbaria. There was a significant controversy at the turn of the century over the classification of the flowering cherry, probably the plant most strongly associated with the Japanese nation. Cherries had a multitude of local cultivars, but in 1901 Jinzō Matsumura (1856–1928) classified the species cherry as *Prunus yedoensis*, claiming his right to classify this national tree.

The early 1900s had seen an enormous consumer enthusiasm in the West for these lovely spring trees, which had previously been limited to arboretum examples. Japanese nursery catalogue illustrations show beautiful images of cherries, made to an extremely high standard at the time. The 1913 collecting trip of Ernest Henry 'Chinese' Wilson was made for exactly this purpose and resulted in his famous *Cherries of Japan* (1916), which is illustrated with black and white photos. However, in this book he quibbled with Matsumura's classification, and instead claimed that the Japanese cherry was a hybrid, not a 'natural' inhabitant of the islands. At the same time a German botanist claimed the cherry actually originated from Korea, not Japan at all. It was frustrating to the Japanese botanists to find their classifications questioned by outsiders. The debate about the cherry shows how the control of the story of botany has been concentrated in Western arboreta, and we shall see in the next chapter some recent moves to address this imbalance.

Finally, towards the end of the Second World War, Willi Hennig (1913–76) began writing his theory of cladistics, published in 1950, which emphasized the need for classification systems to be based on how closely two organisms had a common evolutionary ancestor. Hennig cited the work of Walter Zimmermann (1892–1980), the German botanist who had sought to reconstruct family trees for plants based on the direction that evolution had taken and the pressures driving morphological change. In the next chapter we will see how, post-Second World War, once DNA had been fully understood, a new family tree for the plant kingdom could be created, based on measuring exactly how closely two species were related genetically, opening up stunning new possibilities for botany.

LA VÉGÉTATION DE LA FRANCE, SUISSE ET BELGIQUE, Ire PARTIE

Flore complète
de la France
et de la Suisse

Pour trouver facilement les noms des plantes
SANS MOTS TECHNIQUES

GASTON BONNIER

G. DE LAYENS

5338 Figures
AVEC UNE CARTE DES RÉGIONS DE LA FRANCE
et une carte des régions de la Suisse

PARIS
LIBRAIRIE GÉNÉRALE DE L'ENSEIGNEMENT

TOP AND ABOVE

Gaston Bonnier, French botanist and author of *The Complete Flora of France*.

His flora, which also covered neighbouring Switzerland and Belgium, continues to be updated today and is still referred to as 'Bonnier's'.

RIGHT

Index Plantarum Japonicarum (Index of Japanese plants), (1904)

CENTRE RIGHT

Nippon Shokubutsumeii, (nomenclature of Japanese plants in Latin, Japanese and Chinese), Jinzō Matsumura, (1884)

FAR RIGHT

The Cherries of Japan, EH Wilson, (1916)

The book in which the British botanist Wilson questioned Matsumura's classification.

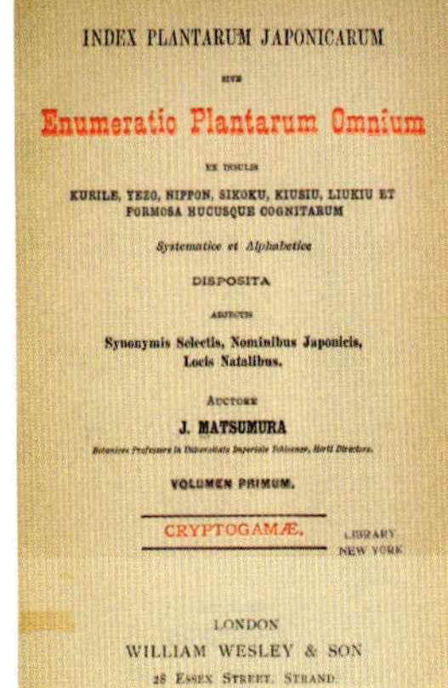
INDEX PLANTARUM JAPONICARUM

SIVE

Enumeratio Plantarum Omnium

EX INSULIS

KURILE, YEZO, NIPPON, SIKOKU, KIUSIU, LIUKIU ET FORMOSA HUCUSQUE COGNITARUM

Systematice et Alphabetice

DISPOSITA

ADJECTIS

Synonymis Selectis, Nominibus Japonicis, Locis Natalibus.

AUCTORE

J. MATSUMURA

VOLUMEN PRIMUM.

CRYPTOGAMÆ.

LONDON
WILLIAM WESLEY & SON
28 ESSEX STREET, STRAND

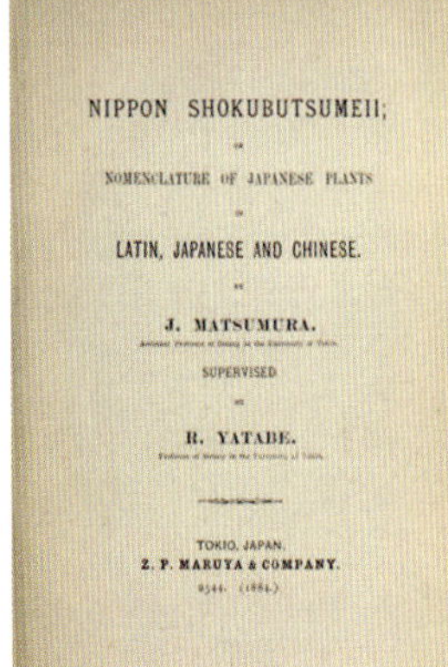
NIPPON SHOKUBUTSUMEII;

NOMENCLATURE OF JAPANESE PLANTS

LATIN, JAPANESE AND CHINESE.

J. MATSUMURA.

SUPERVISED

R. YATABE.

TOKIO, JAPAN.
Z. P. MARUYA & COMPANY.
2544. (1884.)

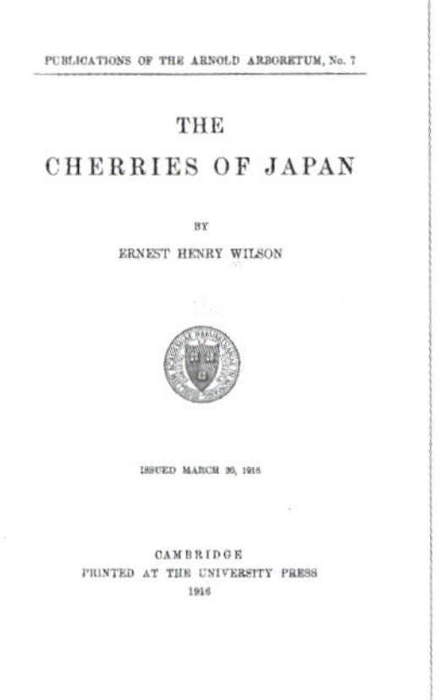
PUBLICATIONS OF THE ARNOLD ARBORETUM, No. 7

THE
CHERRIES OF JAPAN

BY
ERNEST HENRY WILSON

ISSUED MARCH 30, 1916

CAMBRIDGE
PRINTED AT THE UNIVERSITY PRESS
1916

ABOVE

Willi Hennig, the German pioneer of 'cladistic' analysis

Hennig believed understanding their ancestral family trees was the key to creating solid taxonomies for living things.

LEFT

Ernest Lawrence, the American nuclear scientist

Lawrence's lab became a model for research in other areas of science, including botany, after the second world war.

OPPOSITE

Richard Willstätter's Berlin laboratory, 1913

Willstätter was interested in plant pigments, in particular chlorophyll, which he helped to understand as the workhorse of the plant cell.

6

THE MODERN BOTANISTS' LIBRARY TAKES SHAPE

(1950–THE PRESENT DAY)

The modern section of our Botanists' Library, extending from 1950 to the present day, looks very different to the parts covering earlier centuries. The large papyrus scrolls and parchment manuscripts of ancient times are nowhere to be seen, and the extravagantly produced, lavishly illustrated tomes that are ubiquitous in the sixteenth- to nineteenth-century sections are now few and far between. However, there is an extensive collection of journals and textbooks, reflecting the burgeoning array of academic subjects that now fall under plant science's remit, and there are still numerous important floras. Alongside them are many popular science texts, distilling botanical themes into accessible language to engage lay people with the wonder of plants. Increasingly, the library's resources are also available in 'invisible' digital versions, online, to be accessed from inside or outside the library walls, and databases are also now an important part of the botanist's toolset.

That the works in this part of the library seem visually less impressive than those of earlier years belies the vast number of new findings that have emerged from universities and research institutes since the middle of the twentieth century. These advances are mostly reported in journals, many of them dating back to the nineteenth century, where peer-reviewed papers routinely publish new academic research. Numbering more than fifty in the English language alone, they range from *Annals of Botany*, an international publication covering all aspects of plant science, to *Plants, People, Planet*, showcasing plant science relevant to society, and *Plant Physiology and Biochemistry*, which includes contributions across biochemistry, physiology, structure, genetics and interactions between plants and microbes. As we learned in Chapter 4, *Curtis' Botanical Magazine* is the longest running of them all, having begun in 1787. There are also multidisciplinary journals, such as *Nature*, a major academic publication that is renowned for publishing breakthroughs in science, including botany.

1. A discovery takes botany in new directions

It was in 1953 in *Nature* that English biophysicist Francis Crick (1916–2004) and American biologist and geneticist James Watson (1928–) published a groundbreaking paper called 'Molecular Structure of Nucleic Acids: A structure for Deoxyribose Nucleic Acid' (DNA). DNA forms the chemical basis for the information stored in the genes of all living things, shaping how organisms grow, develop, function and reproduce. Following the discovery of the molecule by Swiss physician and biologist Friedrich Miescher (1844–95) in the late 1860s, scientists had slowly identified DNA's key chemical components and worked out how they were joined to each other. Then, in 1952, British chemist Rosalind Franklin (1920–1958) and her PhD student Raymond Gosling (1926–2015)obtained sharp X-ray diffraction photographs of DNA. These enabled Crick and Watson to ascertain that DNA's structure was a three-dimensional double helix, comprising entwined spirals joined to one another by a pattern of amino acids. Scientists could now see how genes could hold information, copy themselves and pass on the information to future generations, revolutionizing understanding of evolution and heredity. Hailed as among the most significant scientific breakthroughs of the twentieth century, it underpinned advances in many aspects of botany. Specifically, it provided tools and insights that helped scientists to better understand plants' biology, hone systems of taxonomy and classification, reconstruct species' evolutionary histories and breed plants with desirable traits.

LEFT

The Botanical Magazine; Or Flower-Garden Displayed, (1790)

Echinacea purpurea by Sydenham Edwards, as featured in an early edition of the magazine. Now known as *Curtis' Botanical Magazine* after its first Editor William Curtis, it is the world's longest-running continuously published botanical magazine. Today published by Wiley, it remains an important forum for botanists and scholars interested in botanical illustration.

At around the time that DNA was revealing its secrets, a few scientists were realizing that plant physiology was so complex, it might be wise for researchers to apply their different specialities to a model organism. With all flowering plants closely related, such a plant could then stand as a proxy for the others. The thale cress (*Arabidopsis thaliana*) was a suitable candidate thanks to its fast lifecycle, close relationship to important crops, ability to grow in a laboratory and small genome (the complete set of genes in a cell or organism). From the 1980s, the scientific community increasingly used *A. thaliana* to integrate classical disciplines of plant science with the expanding fields of molecular biology and genetics. Then in 2000, *A. thaliana*'s whole genome was sequenced – the first flowering plant to gain this accolade – with the research published in another *Nature* paper: 'Analysis of the genome sequence of the flowering plant *Arabidopsis thaliana*'. This diminutive plant, to which Linnaeus had given the specific epithet 'thaliana' following Johannes Thal's description of it in 1588 from the Harz mountains, and of which William Curtis reported in his *Flora Londinensis* (*Flora of London*) of 1777 'no particular virtues or uses are ascribed to it', was set to become one of the most useful plants in botany's history.

Studies of *A. thaliana* threw open a window onto the molecular, cellular and developmental mechanisms that underpin plant life. They helped scientists to understand: how plants develop, such as by putting out embryonic shoots and roots; the mechanisms via which plants photosynthesize and fix carbon; how plants respond to environmental cues, such as light and water; the roles hormones play as plants grow, develop and respond to stress and how genes regulate such events; and epigenetic mechanisms, influencing gene expression. The research even gave insights into how plants interact with microbes, including symbiotic and pathogenic relationships. By 2015, more than 50,000 research papers had been published about the plant, with findings also disseminated to the *Arabidopsis* community of researchers via *The Arabidopsis Book*, a free-access, peer-reviewed serial publication published between 2002 and 2019 by the American Society of Plant Biologists. The work informed diverse disciplines, from agriculture to medicine. Today, better molecular biology and sequencing tools, and the need to study the genetic basis for unique processes in different species, means other plants are also moving into the botanical spotlight, but *Arabidopsis* research nonetheless remains important.

2 The ongoing task of unravelling plant relationships

Findings from new plant-science studies fed into ongoing efforts to classify the world's flora. After the Second World War, this taxonomic work was still the preserve of individual or small groups of botanists, with systems developed by the Soviet-Armenian botanist Armen Takhtajan (1910–2009) and the American botanist Arthur Cronquist (1919–92) becoming widely used. Takhtajan's system drew on the latest insights emerging from studies in plant forms (morphology), embryology, plant cell structure and function, molecular biology and pollen analysis to divide flowering plants into two classes: Liliopsida (Monocotyledons) and Magnoliopsida (Dicotyledons). He published his work in *Systema Magnoliophytorum* (1987), *Diversity and Classification of Flowering Plants* (1997) and, the culmination of his sixty years of study, *Flowering Plants* (2009). Cronquist modified Takhtajan's system, and drew on the work of earlier Charles E. Bessey (1845–1915), publishing his work as a series of texts and monographs in *The Evolution and Classification of Flowering Plants* (1968) and *An Integrated System of Classification of Flowering Plants* (1981). The 'ownership' of such classification systems by individuals – albeit those with considerable authority – often led to heated debates within the community.

While Cronquist and Takhtajan were each drawing together the strands of their life's work, thinking around plant classification was shifting. A new field was developing – systematics – based around the idea that taxonomic classifications should reflect genealogical relationships rather than be based on physical similarities. It was not the first time this idea had been mooted – in *On the Origin of Species* Charles Darwin had said 'our classifications will come to be, as far as they can be so made, genealogies'. Joseph Hooker, too, considered the concept to be important. But it was only in the mid-twentieth century that two concrete ideas emerged as to how this might be achieved.

FAR LEFT

***Sylva Harcynia*, Johannes Thal, (1588)**

This woodcut is the earliest image of the plant we now known as *Arabidopsis thaliana*, but which Thal named *'Pilosella siliquata'*.

LEFT

***Flora Londinensis*, William Curtis, (1777–1798)**

Arabidopsis thaliana was among plants recorded in this six-volume work as growing 'wild in the environs of London' in the late 18th century.

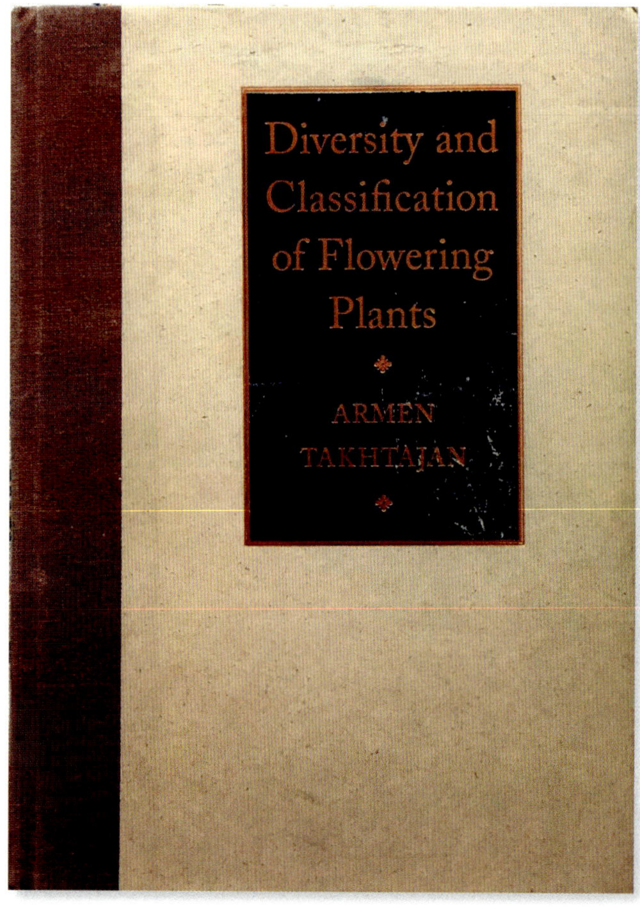

ABOVE AND RIGHT

Armen Takhtajan, (1910–2009)

A Soviet-Armenian botanist, Takhtajan made an important contribution to 20th-century plant evolution and systematics, through his work *Diversity and Classification of Flowering Plants*.

Peter Sneath (1923–2011) and Robert R. Sokal (1926–2012) proposed to use objective, quantitative and numerous features to determine similarity between organisms, outlining their approach in their 1973 work *Numerical Taxonomy*. Meanwhile, the German zoologist Willi Hennig (1913–76) favoured using 'sister groups'. He considered that two taxa (any taxonomic group, such as species or family) were sisters if they were more related to each other than a third group – as evidenced by features shared across the sister groups but absent in the other group. Hennig's book, *Phylogenetic Systematics* published in 1966, outlined how to construct phylogenetic trees and use their branching patterns as the basis for classifications.

Pairing Hennig's theories with DNA sequencing was to revolutionize understanding of the relationship between species. The starting point was the 1998 paper 'An ordinal

ABOVE AND LEFT

Arthur Cronquist, (1919–1992)

Cronquist was highly influential in botany with his 20th-century work *An Integrated System of Classification of Flowering Plants.*

classification for the families of flowering plants' published in the journal *Annals of the Missouri Botanical Garden*. This reported the results of the analysis of 462 flowering plants (angiosperms) based on DNA sequences of a major gene used in photosynthesis. It was a collaborative effort between more than forty scientists who had collected sequences for the same genes, so they could be combined. The results confirmed some expected relationships between species but also threw up some surprises. The 'monocots', such as grasses, lilies and orchids, appeared as a group, for example, but the 'dicots', including magnolias, roses, laurels and daisies, did not. Rather, the monocots now appeared to be one of a number of defined groups within the angiosperms.

The revelation that the sacred lotus (*Nelumbo nucifera*) was more closely related to plane (*Platanus*) trees and banksias, than the waterlilies (*Nymphaea*) that it closely

BELOW

John Reeves Collection, Natural History Museum (late 18th–early 19th century)

DNA evidence showed that the sacred lotus (*Nelumbo nucifera*), depicted here, is more closely related to plane trees than waterlilies.

resembled, was such a shock to the researchers that they re-sequenced the species' gene. But the results came back the same. After tests of equivalent datasets for other DNA regions showed the same picture as the 1993 study, scientists felt sufficiently confident to develop a new classification system, in 1998, based on DNA sequences. They named it the Angiosperm Phylogeny Group (APG) classification, reflecting the wider group of scientists involved. The classification was revised in 2003, 2009 and 2016, as more data became available.

Many botanic gardens and herbaria subsequently adopted the APG system. Doing so was quite an upheaval, requiring the reorganization of specimen filing systems in herbaria, and the replanting of order beds – the formal arrangements of plants used to teach students in botanic gardens. Updates were needed, too, to publications focussed on classifying particular families of plants, including those covering the palms.

ABOVE

Genera palmarum (2008)

Over the past three decades, the classification of palms has progressed rapidly.

When *Genera Palmarum: The Evolution and Classification of Palms* was first published in 1987, this comprehensive reference book included 200 genera and 2,700 species of palms. The second edition, published two decades later, classified 2,500 species in 183 genera. Advances in the use of genetic tools for classifying plants had revealed species initially considered distinct to be variants of other species, with entirely new species also incorporated. The latter included *Tahina spectabilis*, a very large palm with fan-shaped leaves spanning 5m (16ft) from Madagascar, which represented not only a new species but a new genus too. By 2016, the figures had shifted again, with two of the *Genera Palmarum* authors, William Baker (1972–) and John Dransfield (1945–), reporting updated figures of 181 genera and 2,600 species in an article in the *Botanical Journal of the Linnean Society*. The island of New Guinea proved to be particularly rich in palm species, with the newest palm book from Baker and co-authors, *Palms of New Guinea* (2024), covering 250 species. Of these, ninety-one were described as new to science during the book's decades-long creation.

In 2007, the genus Banksia had more than doubled in size when DNA research indicated that it should subsume ninety-four species formerly considered to reside in the genus *Dryandra*. While many Australian herbaria followed the change, *Banksia* expert Alex George (1939–) was against it, believing that plant classification should be 'practical, usable by informed but not necessarily expert users' and therefore should continue to be based in 'readily observable morphological characters'. Among other works, George had written the monograph *The genus Banksia L.f. (Proteaceae)* in 1981, the first systematic reclassification of the genus since George Bentham's *Flora Australia* of the 1870s, and he had also contributed texts for botanical artist Celia Rosser's three-volume tome *The Banksias* (1981). Rosser (1930–) had taken twenty-five years to illustrate all the known species of *Banksia* for this work, which, when the third volume was published in 2000

BELOW

Celia Rosser, photographed by Wayne Taylor

Australian artist Celia Rosser holds aloft a cutting of *Banksia Rosserae*, which was named after her. Celia spent 25 years illustrating all the *Banksia* species only for the genus to be greatly expanded on the basis of new DNA evidence.

numbered seventy-six. It was the first time a genus had been captured in its entirety. The experience shows that plant classification remains fluid, and that even in the days of collaborative classification, debates over taxonomy continue.

3 Floras continue to define botany of specific places

As outlined in earlier chapters, floras have long been the vehicle for resolving uncertainties over what plant names should apply to what species. The second half of the twentieth century saw the development of some notable large-scale projects, including: *Flora Europaea* (*Flora of Europe*), combining national floras of European countries in a single volume (published between 1964 and 1993); the 22,000-page *Flora of the USSR* (initiated in 1931 and published in 1964); and *Flora Reipublicae Popularis Sinicae* (*Flora of the People's Republic of China*), a forty-five-year programme of documentation undertaken in China from 1959. The latter was one of the largest such projects in the world, involving four generations of Chinese plant taxonomists, and covering more than 31,000 species, over half of which only grow in China.

The *Flora of Australia* project was approved in 1979 (a feat that, in itself, had taken twenty years), with the first of sixty planned volumes published in 1981–82. And the *Flora of North America*, rooted in an earlier terminated project, was first mooted around this time, and finally completed in 2024 in thirty volumes. Numerous large-scale serial floras covering parts of the tropics were also executed from the mid-twentieth century. These include the *Flora of Tropical East Africa*, documenting all the known wild plants of Tanzania, Uganda and Kenya. This was the largest regional tropical flora ever created – covering 3–4 per cent of known plant species in the world; work commenced in the 1940s and was completed in 2012.

BELOW

Flora Europaea
various authors,
(1964–1993)

This compendium of the plants of Europe enabled anyone to name flowering plants, ferns or conifers to subspecies level.

Advances in computing technology from the latter part of the twentieth century have hugely influenced how botanical information is made available, who can access it and how it can be queried and used. While floras remain as important as ever today, there has been a shift towards making them available in digital formats. For example, a digitized version of the *Flora Europaea* is now combined with *Med-Checklist*, *Flora of Macaronesia* and other sources as part of the Euro+Med PlantBase, accessible via a data portal. In 2018, *Flora of Australia* was transitioned from the hard-copy book series to an online platform. And *Flora of North America* is being made available online as well as in print. With 1,100 of the featured plants in the latter flora only described in the past twenty years, the benefits of an easily updatable online system are clear. Meanwhile, the *Flora of West Tropical Africa*, initially produced between 1954 and 1972, was reproduced in 2014 in digital e-book format, the first tropical regional flora to be made available in this way. Instead of having to carry the three-volume publication into the grasslands and forests of West Africa, taxonomists could now access it on their mobile phones.

The herbaria sheets that underpin many floras are also being

THIS PAGE AND OPPOSITE

Flora of the People's Republic of China, **(1959–2004)**

One of the largest floristic projects in the world, the final work comprised 80 volumes, in 126 parts, written across more than 40,000 pages.

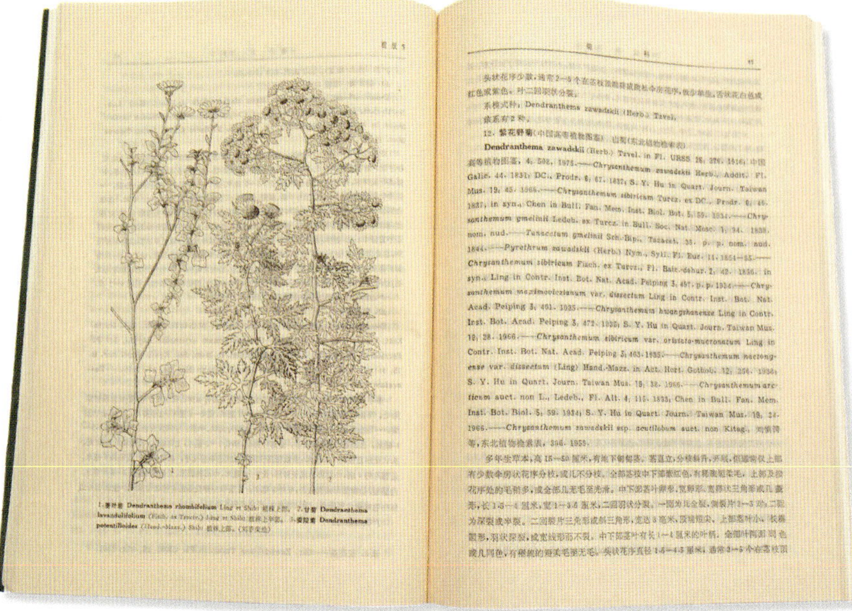

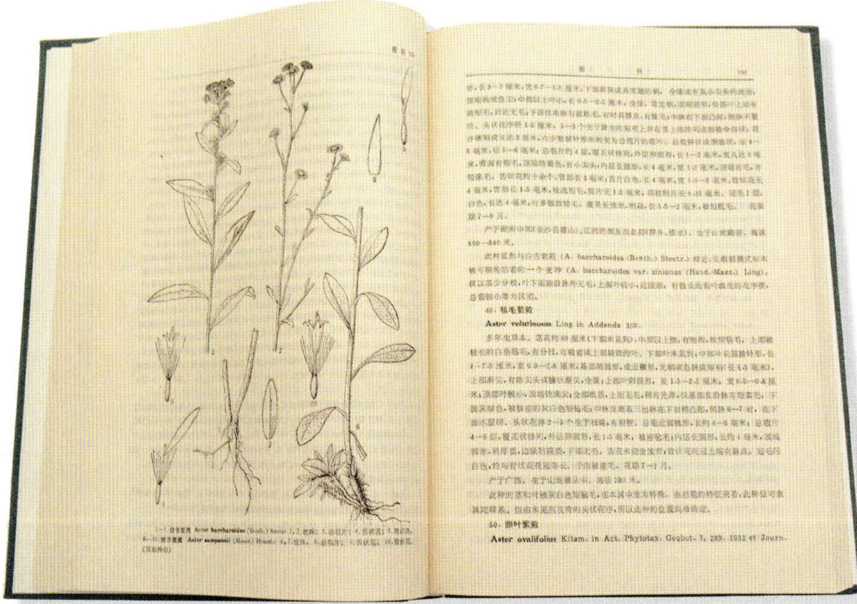

ABOVE

***Reflora Virtual Herbarium,* Multiple contributors, (Launched 2006)**

Making digital scans of Brazilian plants preserved in European herbaria is helping to ensure scientists in Brazil can research their nation's flora.

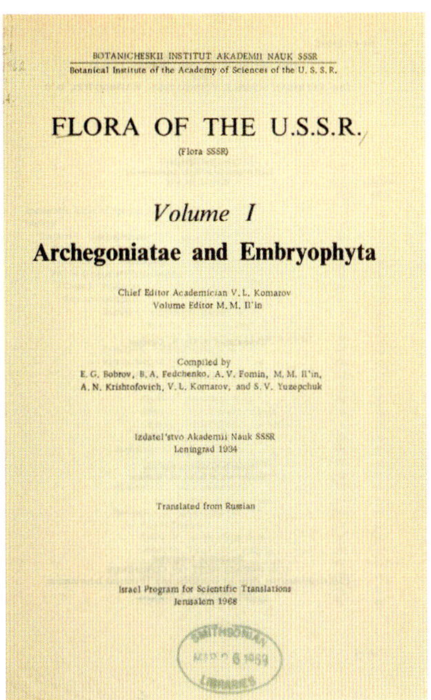

BOTANICHESKII INSTITUT AKADEMII NAUK SSSR
Botanical Institute of the Academy of Sciences of the U. S. S. R.

FLORA OF THE U.S.S.R.

(Flora SSSR)

Volume I

Archegoniatae and Embryophyta

Chief Editor Academician V. L. Komarov
Volume Editor M. M. Il'in

Compiled by
E. G. Bobrov, B. A. Fedchenko, A. V. Fomin, M. M. Il'in,
A. N. Krishtofovich, V. L. Komarov, and S. V. Yuzepchuk

Izdatel'stvo Akademii Nauk SSSR
Leningrad 1934

Translated from Russian

Israel Program for Scientific Translations
Jerusalem 1968

ABOVE LEFT

Flora of the U.S.S.R.
various authors,
(1964)

This flora arose from one of several major projects to document nations' and regions' plants during the 19th century.

digitized to facilitate wider access to specimens: the Reflora project coordinated by the Brazilian National Council for Science and Technology is one example. The hope is to 'virtually repatriate' specimens gathered from Brazil in the eighteenth, nineteenth and twentieth centuries that today reside in major European botanical institutions. This will help to overcome inequality of access to data arising from the colonial period, and enable Brazilian and other scientists to use this vital botanical resource.

4 The poor state of the environment comes into focus

The continued acquisition of basic data about plants and the wider environment, coupled with the digital revolution, enabled conservationists and scientists to better gauge the health of the planet from the late twentieth century. This led to increasing awareness that both the Earth's climate and biodiversity were being threatened by human activities. In 1975, American geochemist Wallace Broecker (1931–2019) introduced the term 'global warming' in his article 'Are we on the brink of a pronounced global warming' published in *Science* magazine. By 1990, the *First Assessment Report of the Intergovernmental Panel on Climate Change (IPCC)* was describing climate change as a challenge with global consequences, and predicting potential future climate changes and impacts. The Millennium Ecosystem Assessment, meanwhile, carried out between 2001 and 2005 by 1,360 experts around the world, found that in the preceding fifty years, humans had changed ecosystems more rapidly and extensively than in any comparable period of time in human history. And the fifth IPCC assessment, of 2014, reported significant warming to Earth's climate system and stated with high confidence that human activities, particularly the emission of greenhouse gases, were the dominant cause of warming since the mid-twentieth century.

Two years later, Kew Gardens, launched its first 'State of the World's Plants' report, providing, for the first time, a baseline assessment of knowledge on the diversity of plants on Earth, the global threats facing them and policies protecting them. The latter included the Convention on International Trade in Endangered Species of Wild Fauna and Flora (CITES), aimed at ensuring international trade in animals and plants did not threaten their survival, and the Nagoya Protocol, a supplementary agreement to the 1992 Convention on Biological Diversity, seeking to ensure that the benefits arising from the use of genetic resources were shared in a fair and equitable way. The days of explorers arriving on foreign shores and ruthlessly helping themselves to plants were gone, but the harm arising from the long shadow of human exploitation of the environment was coming into clear view. The fourth report found that two in five plants were threatened with extinction. It also considered fungi alongside plants (fungi having been classified as a separate kingdom from 1969 but still frequently studied under botanical organizations). It focussed on the potential to use plant and fungal resources better to provide new and sustainable sources of foods, medicines and energy crops.

THIS PAGE

IPCC Assessments, various authors

In 1990, the Intergovernmental Panel on Climate Change began publishing regular scientific assessments of how climate change was progressing and its likely impacts, including on Earth's flora.

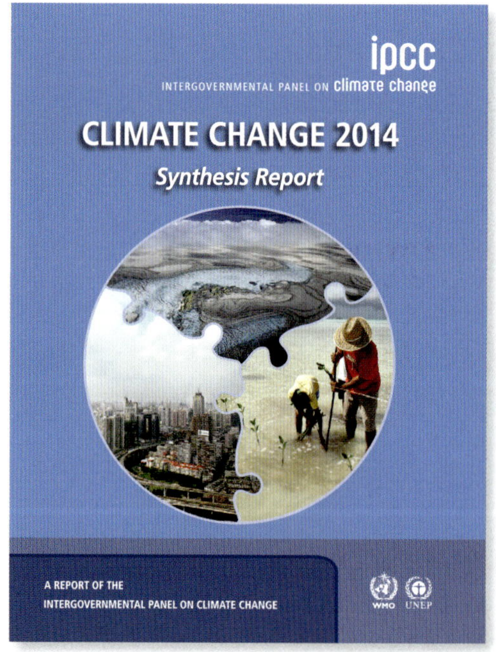

The 'State of the World's Plants and Fungi' (SOTWPF) 2023 report, which involved 200 scientists from 102 institutions in 30 countries, highlighted progress made towards that enduring goal of botanists down the centuries: to know the full extent of the world's flora. Specifically, it focused on the 'World Checklist of Vascular Plants', one of four major inventories of the world's plants, comprising 350,386 entries, with taxonomic and geographic data. It also highlighted that a little determination could go a long way. The 'World Checklist' had been started by Belgian botanist Rafaël Govaerts (1968–), who, while a student in the late 1980s, wanted to identify plants that might be threatened by rainforest destruction. On asking botanic gardens in Brussels for a list of all the world's plants he was told no such list existed and decided to create one. This task, which involved identifying all the evidence that formed the basis for a plant taxon (such as a species, genus, family and so on) and confirming the 'accepted' name and any synonyms, took him three decades. Information on the distribution of each plant was later added, facilitating powerful geographic analysis. Charles Darwin, who a century earlier had sought a similar list, indirectly helped to fund the work through a bequest he left to Kew Gardens. The money he had donated supported the creation of *Index Kewensis*, a list of seed plants maintained by the Gardens from 1885, which later provided the foundation for the International Plant Names Index – on which the 'World Checklist' was based.

THIS PAGE

***State of the World's Plants and Fungi* reports**

Since 2016, the Royal Botanic Gardens, Kew, UK, has published regular evidence-based digests outlining current knowledge on plants and fungi around the world.

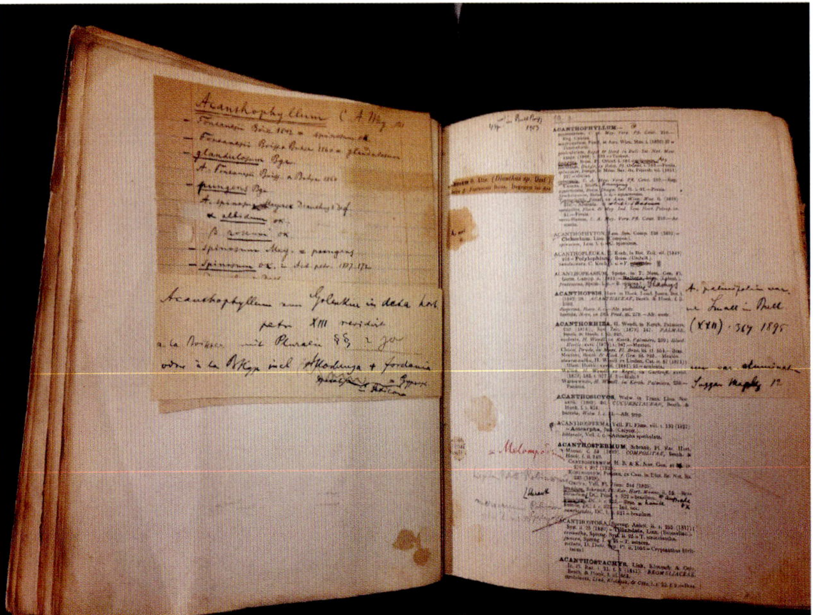

ABOVE

Index Kewensis
(from 1885)

This list of seed plants maintained by the Royal Botanic Gardens, Kew, UK, gave rise to the *World Checklist of Vascular Plants*, one of four current major plant inventories.

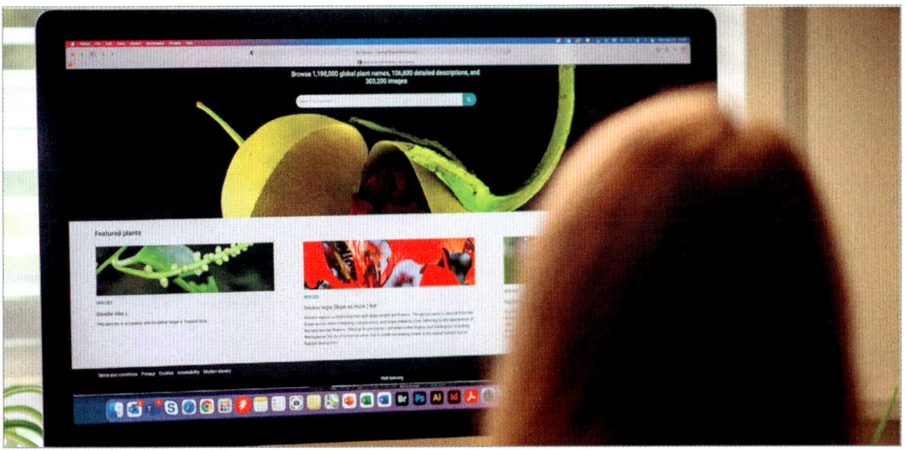

ABOVE

Plants of the World Online

Anyone can look up a plant that is on the World Checklist of Vascular Plants - through the portal *Plants of the World Online.*

SOTWPF 2023 demonstrated the power of combining the taxonomic and geographic information of the 'World Checklist' with genetic and other data. It included analysis showing that temperature and rainfall had been important in the evolution of annual-versus-perennial plants. It reported how scientists had been able to build the largest-ever family tree of the orchid family, which showed that orchids dated back eighty-three million years to times when dinosaurs still roamed the Earth. And it demonstrated that more than 220,000 plant species are endemic – unique to a single country – and are therefore at the mercy of that nation's conservation policies.

A problem, however, was that having a complete list of all the world's plants was still a long way off: the report also estimated that 15 per cent of the world's flora – some 50,000 species – had yet to be located, and officially described and named. With the 2019 'Global Assessment Report on Biodiversity and Ecosystem Services' of the Intergovernmental Science-Policy Platform on Biodiversity and Ecosystem Services declaring that 'Nature and its vital contributions to people, which together embody biodiversity and ecosystem functions and services, are deteriorating worldwide,' time is running out to find them.

Knowledge about the general health of individual species largely comes from the Red List of Threatened Species, the go-to resource for information on the conservation status of animals, fungi and plants. First published by the International Union for the Conservation of Nature (IUCN) in 1964, it categorizes species as Extinct, Extinct in the Wild, Critically Endangered, Endangered, Vulnerable, Near Threatened, Least Concern or Data Deficient, either globally or within a particular country or region. This data is important for informing conservation policy. However, to date, only 157,000 known species of plants have been assessed for the IUCN Red List, and three-quarters of as-yet unnamed plants are already likely to be threatened with extinction. The situation is even worse for fungi: the SOTWPF 2023 report found that between 92 per cent and 95 per cent of fungi had yet to be named. Rapid triage, using Artificial Intelligence, is helping plant scientists to identify priorities for conservation faster.

5 Illustration remains vital to plant science

You might think that the high-tech turn plant science has taken in recent decades would have rendered botanical art obsolete. But that is far from the case. It remains a fundamental part of classification because scientific illustrators are able to portray individual plant parts with ease. Often, specimens collected in the field are dried and pressed to preserve them – and it is these that taxonomists use to describe and classify plants. On examining a specimen, a botanical artist can emphasize parts that are critical for classification, making it simpler for the taxonomist to compare and contrast the specimen with others – and either classify the plant as a known species or determine that it is new to science. While photography can also be useful within plant science – such as to record a plant's colours – the camera cannot recreate a lifelike botanical specimen from a two-dimensional preserved herbarium specimen in the way an artist can.

Sometimes, botanical art can even be a force for finding new species. The 2022 discovery of *Victoria boliviana* – the largest waterlily in the world, with 3m (10ft) wide leaves – was stimulated after artist Lucy T. Smith began a personal project to illustrate, life-size, all the species of waterlily. She was able to demonstrate to experts that one of the giant waterlilies she was working on was different to the two known giant waterlilies, *Victoria amazonica* and *Victoria cruziana*. A specimen of the new species had been sitting in Kew Gardens' herbarium for 177 years before Smith's work prompted the major taxonomic investigation that revealed the new species. And when the academic paper 'Revised species delimitation in the giant waterlily genus *Victoria* (Nymphaeaceae) confirms a new species and has implications for its conservation' was published in 2022

RIGHT

Victoriana boliviana

Despite its huge size, this waterlily was only named as new to science in 2022, a discovery prompted by the keen observations of a botanical illustrator.

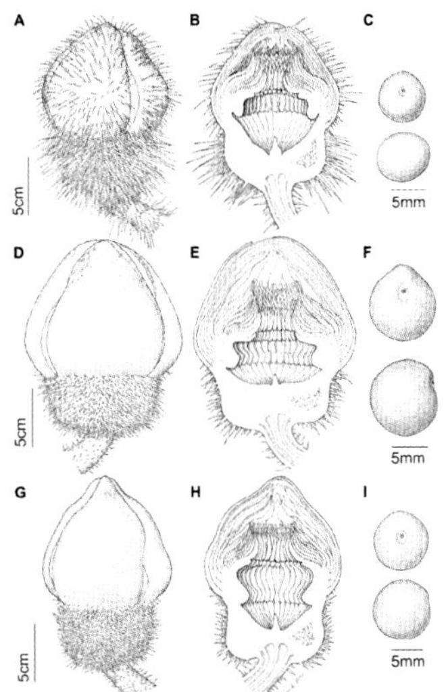

THIS PAGE

***Frontiers in Plant Science*, various authors,**
(2022)

These images hone in on details that distinguish *Victoria boliviana* from the two other giant waterlilies in the genus.

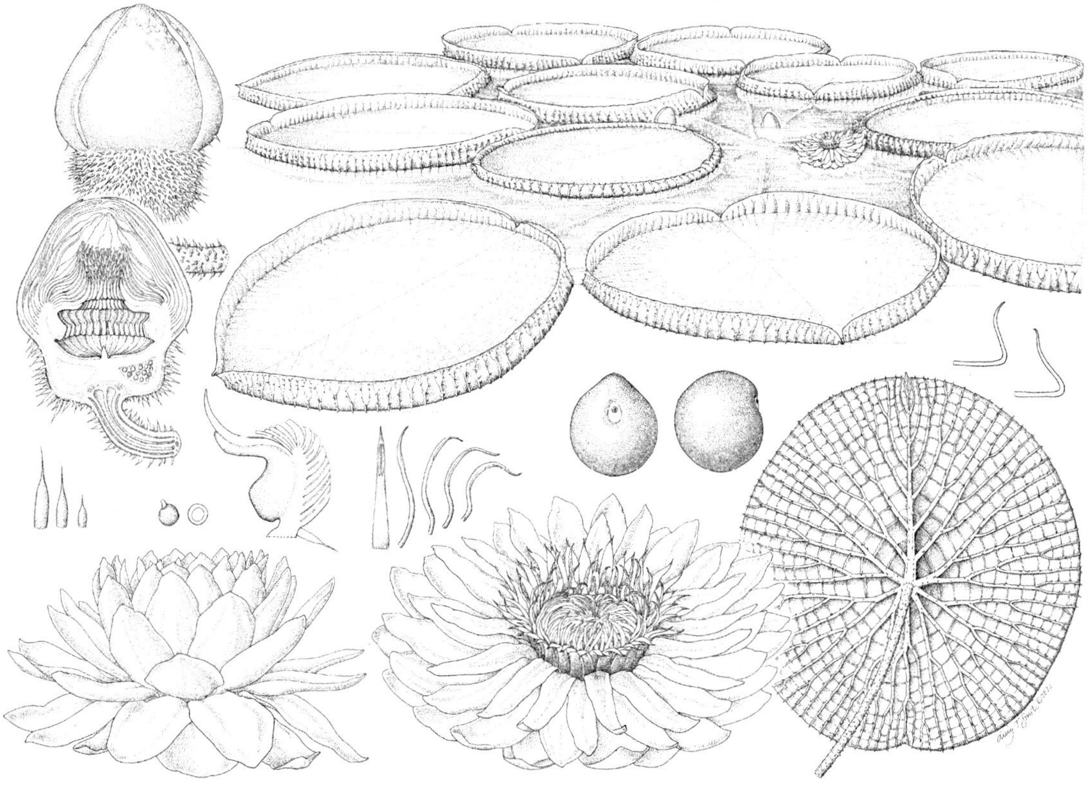

ABOVE
The Florilegium, (2018)

The works of sixty-four botanical artists are presented in this celebration of plants growing in the Royal Botanic Gardens, Sydney, Australia.

in the journal *Frontiers in Plant Science*, Smith's illustrations were there to clearly show readers the evidence for the decision. Meanwhile, the fruits of her personal project were displayed at the Shirley Sherwood Gallery at Kew Gardens, the world's first gallery devoted to classical and contemporary botanical art. The brainchild of botanist and art collector Shirley Sherwood, it displays images from Sherwood's personal collection – also showcased in the book *The Shirley Sherwood Collection* – as well as those of Kew Gardens.

Large, lavish volumes illuminating the beauty of particular collections of plants have been less numerous in the modern period than earlier, but one such work stands out. That is *The Highgrove Florilegium*, produced in 2016, for which seventy-two leading botanical artists from around the world were asked to depict selected plants from the garden of HRH The Prince of Wales and the Duchess of Cornwall (now HM King Charles III and HM the Queen Consort) at Highgrove, Gloucestershire, England. The work encompasses 124 plates, showing plants from showy *Rhododendron* to delicate *Cistus*, accompanied by texts that name the plants in Latin and describe their habitats. Produced in 175 numbered sets of two double-elephant folio volumes, the work sells for more than £12,000, with the royalties going to charity. The then Prince also commissioned the 2018 work *The Transylvania Florilegium*, in which thirty-six botanical artists recorded wild plants such as campanula, crocus and geranium that grow in this part of Romania.

A number of florilegium societies exist, which are concerned with documenting specific collections of plants. Some artworks commissioned by such societies have been published in books, with others displayed in exhibitions or online. Published works include *A Florilegium: Sheffield's Hidden Garden*, produced by the Florilegium Society at Sheffield Botanical Gardens and Oxley, UK (2021); *The Florilegium: The Royal Botanic Gardens Sydney: Celebrating 200 Years*, produced in Australia (2018); and the *Alcatraz Island Florilegium*, containing artworks by members of the Northern California Society of Botanical Artists. The latter has a unique history. Initially built in 1850 as a fort on the Isla de los Alcatraces (Island of the Pelicans), a rugged rock off San Francisco, California, US, Alcatraz was a maximum-security federal prison for three decades from the 1930s, during which time Victorian-style gardens were created there. When the prison closed in 1963, the gardens were abandoned for forty years but restored in 2003. Today, an array of plants thrives in them including roses, fig trees, bulbs and succulents. Featured in the online version of the florilegium are the plume albizia (*Albizia distachya*), by Sally Petru, Dahlia 'Thomas Edison' by Stephanie Rozzo, and foxglove *Digitalis purpurea* 'Camelot Rose' by Susan Hill-McEntee.

Increasingly, botanical illustrations are being used as a focus for 'decolonizing' collections and unearthing new narratives from the past. In particular, work is under way to name the Indian artists who created beautiful illustrations of plants for the British East India Company (EIC) in the eighteenth and nineteenth centuries – but whose identities

LEFT

The Transylvania Florilegium, (2018)

This lavish volume of plant portraits showcases the flora of central Romania.

BELOW LEFT

The Shirley Sherwood Collection, Shirley Sherwood, (2019)

The Shirley Sherwood Collection of contemporary botanical art, showcased in this book, is considered the most important private collection of its kind in the world.

LEFT

The Alcatraz Florilegium, (2016)

This book showcases images of plants growing in the gardens of the former high-security prison on Alcatraz Island, USA.

BELOW LEFT

The Alcatraz Florilegium, (2016)

Plants such as this foxglove (*Digitalis purpurea* 'Camelot rose') by Susan Hill-McEntee once brightened up the lives of prisoners and officers on Alcatraz.

BELOW RIGHT

The Alcatraz Florilegium, (2016)

Plume albizia (*Albizia distachya*) by Susan Petru. Native to Indonesia and southwest Australia, this small tree thrives in Alcatraz's subtropical desert climate.

LEFT

The Alcatraz Florilegium, **(2016)**

The eye-catching garden favourite Dahlia 'Thomas Edison', captured in fine detail by Stephanie Rozzo.

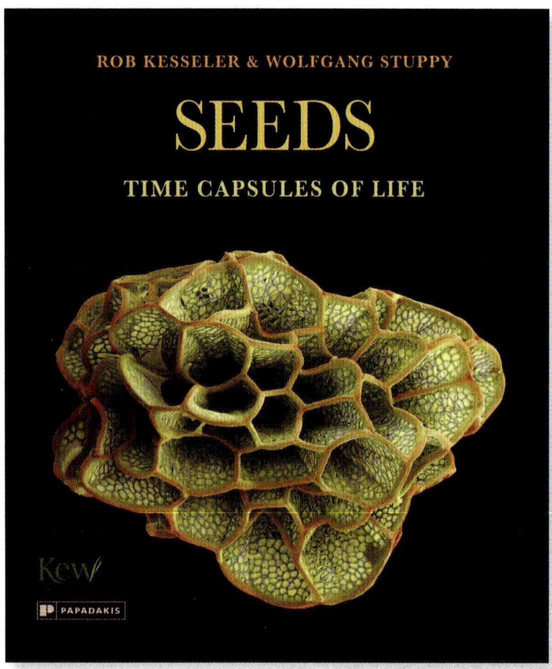

ABOVE
Seeds: Time Capsules of Life, Robert Kessler and Wolfgang Stuppy, (2024 edition)

Science and art collide in this portfolio of seeds captured by scanning electron microscope technology.

were lost by application of the umbrella term 'Company School' or '*Kampani kalam*'. This title is now questioned not only due to its focus on the European EIC commissioners of the artworks but also for concealing artistic diversity among the artists. Henry Noltie (1957–) of the Royal Botanic Garden Edinburgh has been reuniting disparate but connected paintings, herbarium specimens and manuscript archives from the period, and using evidence contained within such records at both Edinburgh and Kew Gardens, to identify the individual artists and tell their hidden histories. Using this method he has been able to, for example, identify paintings by the artist Vishnu Prasad, who worked at Calcutta Botanic Garden (now the Acharya Jagadish Chandra Bose Indian Botanic Garden) in the nineteenth century. The book *Indian Botanical Art: An Illustrated History* by horticulturalist, author and former editor of *Curtis' Botanical Magazine* Martyn Rix, charts progress in this regard, and showcases the works of these unsung Indian artists.

While historic botanical illustration is revealing new stories from the past, a novel form of botanical imagery has emerged from the melding of the modern technology of scanning electron microscopy (SEM) with science and art. The development of SEM from the 1920s helped to revolutionize plant science from the mid-twentieth century, by enabling researchers to visualize cell and tissue morphology at much higher magnifications than was possible using traditional light microscopes. For example, SEM illuminated in fine detail leaf surfaces, including pores or 'stomata' and plants' tissues, which led to better understanding of plant anatomy and how plants interact with the environment and pathogens. It also presented opportunities to showcase the beauty inherent in the magnified parts of plants, including seeds. One book that exemplifies this is *Seeds: Time Capsules of Life* by artist Rob Kesseler (1951–) and seed morphologist Wolfgang Stuppy (1966–), first published in 2012, which provides a natural history of seeds illustrated with close-up photographs and scanning electron micrographs.

6 Shaping future generations of plant scientists

The great strides made in plant science in the decades since 1950 has necessitated publication of an increasing number of textbooks to educate the next generation of botanists. Covering topics spanning plant physiology, ecology, taxonomy, genetics, anatomy and morphology, pathology, biotechnology, evolutionary biology, biochemistry and physiology, there are too many to mention them all. However, several works are worth noting for their enduring appeal. An example is the 1953 book *Plant Anatomy* by German-American plant biologist Katherine Esau (1898–1997). Updated in 1965, a third edition (Esau's *Plant Anatomy: Meristems, Cells, and Tissues of the Plant Body: Their*

Structure, Function, and Development) was written by Ray F. Evert (1931–) in 2006 after Esau's death. Fittingly Evert held the post of Katherine Esau Professor of Botany and Plant Pathology at the University of Wisconsin, USA.

Plant Physiology by Lincoln Taiz (1942–) and Eduardo Zeiger (dates unknown), was first published in 1960 but its appeal was such that it was updated six times, the most recent incarnation being *Plant Physiology and Development* (2023), with Ian Max Møller and Angus Murphy added to the author list. *Ecology* by Eugene Pleasants Odum (1913–2002), which introduced key concepts of ecosystem ecology when it was first published in 1963, exhibited similar longevity. The fifth edition, published in 2004 as *Fundamentals of Ecology* and co-authored by Gary W. Barrett (1940–2022), was the last that Odum was involved with before he died. Odum had also published *Ecology and our Endangered Life Systems* in 1988 (updated to *Ecology: A Bridge between Science and Society* in 1997), which discussed aspects of ecology within the modern context of human mismanagement of the planet. The same year, *Biodiversity* was published, edited by E.O. Wilson (1929–2021), drawing attention to the accelerating loss of plant and animal species due to human-related pressures.

ABOVE

***Flower Portraits*, Joyce Tenneson, (2003)**

Photographers have helped to make the wonder of botany accessible to all.

BELOW

***Gardens of Corfu*, Marianne Majerus, (2018)**

Majerus' images give the reader entrance to romantic old estates, grandmother's plots and contemporary gardens on this Greek island.

7 Bringing botany to a wider audience

Since 1950, a vast number of popular science books have been produced to cater for enthusiasts of plants and botany. While not as learned as many academic texts and floras, they have helped enlighten the public and potential future botanists as to the wonder of plants. Some have simply used photography to showcase plants' beauty. Notable photographers of plants from 1950 onwards include three women, the Americans Imogen Cunningham (1883–1976) and Joyce Tenneson (1945–), and Marianne Majerus (1956–) from Luxembourg. Cunningham worked in black and white, capturing in detail botanical textures and forms, such as those published in the 2001 book *Imogen Cunningham: Flora*. Tenneson has similarly used a limited colour palette, particularly sepia or yellow tones, as showcased in her 2003 work *Flower Portraits: The Life Cycle of Beauty*. Her ethereal plant portraits capture the fragility of life and the ephemeral nature of plant lifestyles. And Majerus, considered one of Europe's leading garden photographers, portrays the colour, shapes and beauty of wider human-shaped landscapes using natural light and shadow. She has been the principal photographer on many books, including *The Great Gardens of London* (2015), *Gardens of Corfu* (2018) and *Gardens of the Italian Lakes* (2016).

In many Western countries, as post-war reconstruction took place and economic prosperity began to grow from the 1950s, new field guides enticed people to venture out into the countryside and identify the wildflowers they found. These books include the 1956 *Collins Pocket Guide to Wild Flowers*, by David McClintock and R.S.R Fitter (boasting over 1,400 illustrations, 600 in colour, all painted from living plants); *The Concise British Flora in Colour*, by W. Keble Martin, first published in 1965, and also beautifully illustrated; and *A Field Guide to Trees and Shrubs* (1958), by George A. Petrides, providing descriptions, illustrations and

RIGHT

Imogen Cunningham (1920s)

Cunningham used black and white imagery to highlight the extraordinary shapes and forms of plants, such as this water hyacinth.

THIS PAGE

Imogen Cunningham billboard (left) and Imogen Cunningham portrait (below)

American photographer Imogen Cunningham is known for her artistic portraits and photographs of botanical subjects, which helped to open people's eyes to the detail and beauty of plants. Here one of her images is used to advertise Art Everywhere US, a vast outdoor art event that took place in the USA in 2014.

THIS PAGE

Marianne Majerus (1956–)

Majerus used natural light to emphasize the beauty of plants in these garden landscapes.

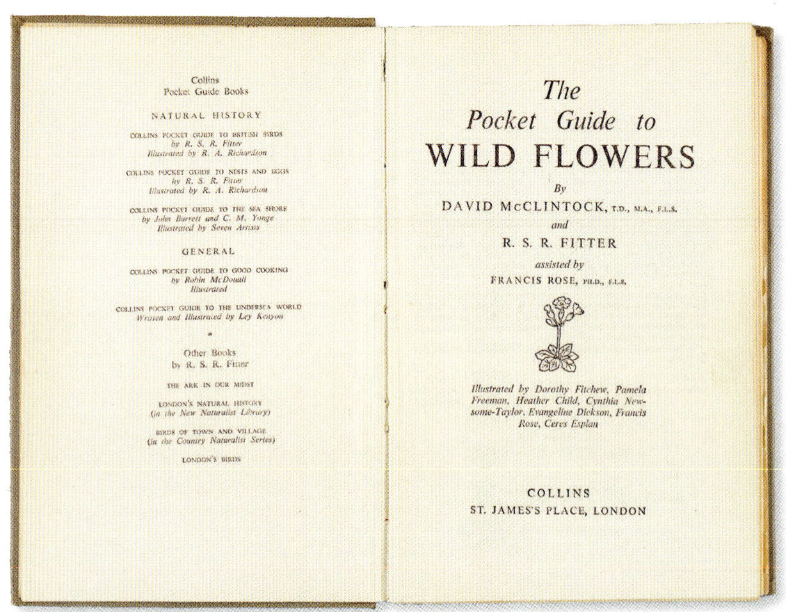

Collins
Pocket Guide Books

NATURAL HISTORY

COLLINS POCKET GUIDE TO BRITISH BIRDS
by R. S. R. Fitter
Illustrated by R. A. Richardson

COLLINS POCKET GUIDE TO NESTS AND EGGS
by R. S. R. Fitter
Illustrated by R. A. Richardson

COLLINS POCKET GUIDE TO THE SEA SHORE
by John Barrett and C. M. Yonge
Illustrated by Seven Artists

GENERAL

COLLINS POCKET GUIDE TO GOOD COOKING
by Robin McDouall
Illustrated

COLLINS POCKET GUIDE TO THE UNDERSEA WORLD
Written and Illustrated by Ley Kenyon

*

Other Books
by R. S. R. Fitter

THE ARK IN OUR MIDST

LONDON'S NATURAL HISTORY
(in the New Naturalist Library)

BIRDS OF TOWN AND VILLAGE
(in the Country Naturalist Series)

LONDON'S BIRDS

The
Pocket Guide to
WILD FLOWERS

By
DAVID McCLINTOCK, T.D., M.A., F.L.S.
and
R. S. R. FITTER
assisted by
FRANCIS ROSE, PH.D., F.L.S.

Illustrated by Dorothy Fitchew, Pamela Freeman, Heather Child, Cynthia Newsome-Taylor, Evangeline Dickson, Francis Rose, Ceres Esplan

COLLINS
ST. JAMES'S PLACE, LONDON

THIS SPREAD

The Pocket Guide to Wild Flowers
David McClintock, (1956)

From the 1950s, an increasing number of field guides encouraged people to explore the flora of their local areas.

BELOW

A Field Guide to Trees and Shrubs
George Petrides,
(1972 edition)

George Petrides' field guide included detailed images to help non-experts identify trees and shrubs.

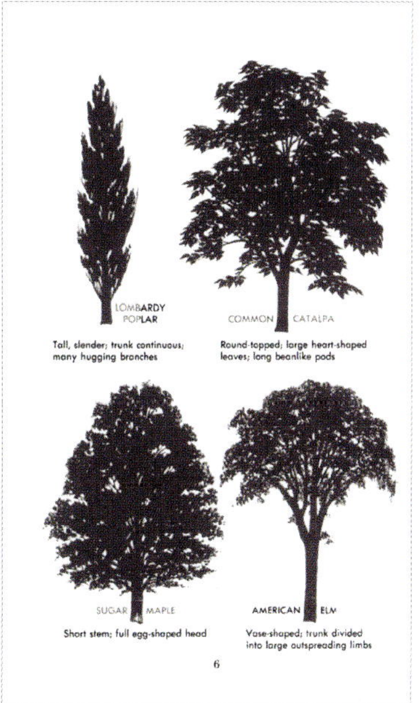

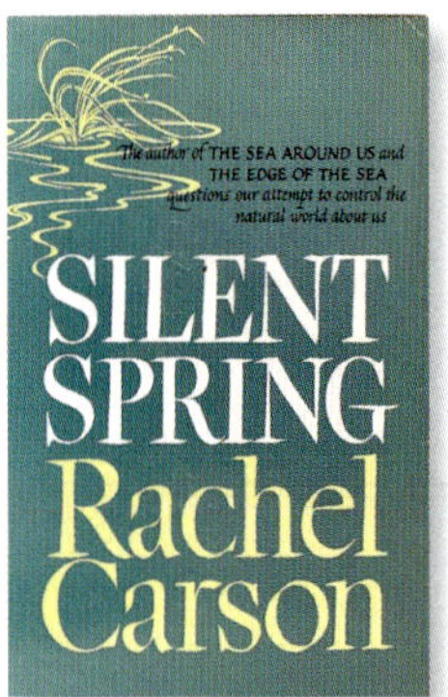

maps to help the novice and expert alike identify trees and shrubs in North America. With motoring in its early days, oil companies sought to encourage people to go exploring in their cars – and in the process, of course, to buy more petrol. One example is the UK *Shell Guide to Flowers of the Countryside*, first published in 1955, which featured beautiful illustrations of British plants by husband-and-wife team Rowland and Edith Hilder. Climate change had yet to make itself apparent, although even back then, scientists knew that increasing atmospheric carbon dioxide, such as by burning fossil fuels for transport, would increase Earth's surface temperature.

It was not long before popular science books began to cover environmental issues. In *Silent Spring*, published in 1962, American marine biologist Rachel Carson (1907–1964) highlighted the dangers of pesticide use on the environment, including plants. In doing so, she helped set the stage for the environmental movement. Subsequently an increasing number of books drew the public's attention to the impacts of human activities – from deforestation to pollution and climate change – on the environment. Among them, botanical artist Margaret Mee's (1909–88) *In Search of Flowers of the Amazon Rainforest*, detailing her travel on foot and by canoe through the forest in the 1950s, 1960s and 1970s, helped to draw attention to the threats from mining to the rainforest. The image that graces the front cover of the book is of *Gustavia augusta*, a small tree native to South America. Instead of being presented on a white background, as is traditional for botanical illustrations, it is set against a dark, shadowy image of its rainforest habitat. Mee took this unconventional approach with many of her paintings, to help raise awareness of the rainforest and the struggles of early conservationists to protect it.

In recent decades, authors have increasingly focussed on the interconnectedness of nature. *A New Look at Life on Earth* (1979) by James Lovelock's (1919–2022) put forward the idea that Earth functions as a single organism. And E.O. Wilson's *The Diversity of Life*, published in 1992, described how the world became diverse in species, how they were interconnected, and why protecting fragile ecosystems should be a priority. The lack of sufficient action to do so meant that by the time Elizabeth Kolbert (1961–) wrote *The Sixth Extinction: An unnatural history* in 2014, she was able t o detail the catastrophic impact of human activities on the environment and explain why it had led to widespread extinction of species. Not all recent popular environmental books related to plants have been downbeat, however. In the 2015 book *Braiding Sweetgrass: Indigenous Wisdom, Scientific Knowledge and the Teachings of Plants*, USA-based botanist and member of the Citizen Potawatomi Nation, Robin Wall Kimmerer, highlights lessons that we can learn from diverse living beings if we are just open to hearing their voices.

8 A visit to the future Botanists' Library

As this chapter shows, today's Botanists' Library has an eclectic mixture of items from diverse authors, aimed at a range of readers. But what of the future library? Certainly, the era in which the scribed, penned or printed work has been the primary means of disseminating information about plants is drawing to a close. While books may well continue to exist as luxury, resource-rich items, botany is truly embracing the digital world, from e-book floras to online checklists of species and globally accessible 'virtual' herbaria. There are even mobile phone apps available today that can identify plants from a quickly snapped photo. In time, it is possible that such an app, underpinned by superior AI and machine learning, might enable anyone, anywhere, to know if they've spotted an unnamed plant. But, in reality, alongside conserving known species, a critical task for future botanists remains to use detailed taxonomic detective work to find and describe the as-yet unknown plants and place them on the global plant family tree. It may seem like a throwback to ancient times but without this fundamental botanical data, the high-tech, globally accessible next-generation Botanists' Library can never be complete.

LEFT

Silent Spring,
Rachel Carson,
(1962)

After Carson, many writers began to illuminate the impacts of human activities on the environment.

BELOW LEFT

***In Search of Flowers of the Amazon Forests*,**
Margaret Mee,
(1988)

Mee used her artworks to highlight the destruction of the Amazon rainforest, an ongoing concern today.

INDEX

Page numbers in *italics* refer to illustrations

A
Achudan, Itty 117
Achun, Win 170
Adanson, Michel, *Familles naturelles des plantes* 176
Adelaide Botanic Garden 212
Africanus, Constantinus 43
Agnivesha 19
al-Biruni, Abu Rayhan Mohammad bin Ahmad, *Kitab al-Saydanah fi al-Tibb* 43
al-Ghafiqi, Abu Jafar Ahmad ibn Mohammad, *Kitab fi al-adwiya al-mufrada* *42*
al-Ghassani, Abul Qasim ibn Mohammed 81
Albert the Great (Albertus Magnus) 43, 111
De vegetabilis et plantis 43
The Alcatraz Island Florilegium 250, *252–3*
Aldrovandi, Ulisse 74, 75, 77, *77*
herbarium catalogue *78*
Alexander the Great 21, *22*, 25
Allen, David 212
Amenhotep I 14
American Society of Plant Biologists, *The Arabidopsis Book* 233
Anderson, Alexander 170
Angiosperm Phylogeny Group (APG) 236
Anicia, Patricia Juliana 26
Annals of Botany 230
Annals of the Missouri Botanical Garden 235
Anne, Queen 77
Anne of Brittany 63
Antón 113
Apuleius 25
Arabidopsis thaliana 232–3, *232*
Aristotelian beliefs 154
Aristotle 21, *21*
Asclepius 76
Atharvaveda 16, *16*, 17
Athenaeus of Naucratis 22
Atkins, Anna 197–8
Photographs of British Algae: Cyanotype Impressions 198, *203*
Attalus III 22
Auge, Johann Andrea 116
Augsburg Bible *55*
Austria 156–60
Ayurveda 6, 16–19, 20, 37, 84
Aztecs *82*, 83, *83*, 113

B
Baker, William:
Genera Palmarum: The Evolution and Classification of Palms 236
Palms of New Guinea 236
Bald's Leechbook 34
Baldung, Hans 64, *64*
Bämler, Johannes 60
Banks, Joseph 137, 140, 141–2, *142*, *143*, *147*, 160, 177, 193
Banksia 236, 239
Barberini, Cardinal 130
Baret, Jeanne 161
Barrett, Gary W., *Fundamentals of Ecology* 255
Bateman, James 197
The Orchidaceae of Mexico and Guatemala 193–6, *193*, *194–5*
Principles of Heredity 213
Bauer, Ferdinand 156–7, 160
Flora Graeca 157, *158–9*
Illustrationes florae Novae Hollandiae 157, 160, *160*
Bauer, Franz 157, 160
Delineations of Exotick Plants, Drawn and coloured and them Botanical characters displayed according to the Linnean System 160
Illustrations of Orchidaceous Plants 193
Bauhin, Casper, *Pinax Theatri Botanica* 106, *106*, 111
Beagle 161
Beale, Mary *131*
Bellairs, Nona 212
Bencao mengquan 34
Benson, Andrew 224–5
Bentham, George 210, *210*
Flora Australia 236
Flora Australiensis 210, *211*
Flora Hongkongensis 210, *211*
Genera plantarum 210, *211*
Bentinck, Margaret 148
Bering, Vitus 154
Besler, Basilius 88, *88*
Hortus Eystettensis 88, *88*, *89*, 91, 100, 127
Bessey, Charles E. 233
Bhatt, Apu 117
Bhatt, Ranga 117
the Bible 33, *33*
Augsburg Bible 55
Bingen, Hildegard von 43, *43*
Physica 43
Blunt, Wilfred 151
Bock, Hieronymus 76–7
Kreutterbuch 76
Boerhaave, Herman 115, 134
Bologna 74
Bonafede, Francesco 74
Bonnier, Gaston *225*
Flore complete illustrée de France, Suisse et Belgique 224
Bonpland, Aimé, *Essai sur la géographie des plantes 162*
Book of Hours 63
Botanic Garden, St Vincent 170
Botanical Journal of the Linnean Society 236
The Botanical Magazine 231
Bougainville, Louis de 160, 161
Voyage autour du monde 160–1
Boussingault, Jean Baptiste 186
Boym, Michał Piotr *127*
Flora sinensis 127, 127–8, *129*
Brahma 16
Brancion, Jean de 77
Brassavola, Antonio Musa 109, 111
Braudel, Fernand 52
Brazil *114*, 115, *242*, 243
Brazilian National Council for Science and Technology 243
British East India Company (EIC) 9, 113, 250, 254
British Ecological Society 216
Broecker, Wallace 243
Brookshaw, George, *Pomona Britannica* 218, *221–3*
Brown, Robert 192, 193
Brunfels, Otto 52, 64, *64*, 76
Herbarium vivae eicones 64, *65*, 66
Bruxella, Arnaldus de 60
Buffon, Georges-Louis Leclerc de 151
Bull, Edith Elizabeth 224
The Herefordshire Pomona 224
Bull, Dr Henry Graves, *The Herefordshire Pomona* 218, 224
Bulliard, Pierre, *Herbier de la France* 151, *155*
Burbank, Luther 213
Burghley, Lord 77
Burman, Johannes
Flora Zeylanica 134, 137
Herbarium amboinense 116
Rariorum Africanarum plantarum 116
Thesaurus Zeylanica 116, *117*, 126
Busbecq, Ogier Ghiselin de 76, 88
Bute, Lord *151*

C
Caesalpino, Andrea 111
Calcutta Botanic Garden 163, *164*, 254
Calvin, Melvin 224–5
Camerarius, Rudolf Jacob 134
Cameron, Margaret *196*
Candolle, Alphonse de 177
Flore Française 177
Prodromus systematis naturalis regni vegetabilis 177

Carson, Rachel,
Silent Spring 262, *262*
Castells, Pieter II 101, *102–5*
Castor, Antonius 25
Catesby, Mark 122, 126
The Natural History of Carolina, Florida and the Bahama Islands 122, *124–5*, *126*
Cathcart, John Ferguson 198, *200*
Cato, Marcus Porcius 25
Cesalpino, Andrea 74, 75
De *plantis libri XVI* 75
Charaka,
Charaka Samhita *18*, 19, 20
Charlemagne 74
Charles IV, King of Spain 162
cherry *222*, 225
Chi Han,
Nan-Fang ts'ao-mu chuang 34, *35*
China 14, 81, 239
knowledge of plants in 34–7, 127–8
materia medica 126–7, 128
printed books 52
Christianity 26, 33, 43
Cibo, Gherardo 74–5
Claudius, Hendrik 116
Clements, Frederic 214, 216
Clevely, John,
View of HMS Resolution and Discovery at anchor in Huaheine 146–7
Clifford, George 134
Clusius, Carolus 84, 88, *88*, 115
Exoticorum libri decem 88
Rariorum plantarum historia 88, 100, *100*
Codex Vindobonensis 26, *26*, *28–31*, 157
Colden, Jane *141*
Columella, Marcus 25
Commerçon, Philibert 161
Company School 9, 250, 254
Compton, Henry 91
Confucius 127
Convention on Biological Diversity 244
Convention on International Trade in Endangered Species of Wild Fauna and Flora (CITES) 244
Cook, Captain James 137, 141, *145*, *147*, 160
Cornwall, Duchess of 250
Crick, Francis 230
Cromwell, Oliver 109
Cronquist, Arthur 233, *235*
The Evolution and Classification of Flowering Plants 233
An Integrated System of Classification of Flowering Plants 233, *235*
Cruz-Badiano Codex 83
Cunningham, Imogen,
Imogen Cunningham: Flora 255, *258*
Curtis, William 142–8
Curtis's Botanical Magazine 148, 231
Flora Londinensis 142, *150*, 232–3, *232*
Curtis's Botanical Magazine 148, *179*, 197, *197*, 198, 230, 231, *231*
Cuvier, Georges 177

D

da Milano, Giovanni,
Regimen sanitates Salernitanum 44
Darwin, Charles 156, 161, 186, 196–7, *196*, 198, *198*, 214, 245
Insectivorous Plants 197
On the Origin of Species 196–7, 233
On the Various Contrivances by Which British and Foreign Orchids Are Fertilised by Insects, and On the Good Effects of Intercrossing 197
The Voyage of the Beagle 196
De materia medica (Dioscorides) 23, 26, *27*, 33, 38
Dickinson, Lowes *210*
Diderot, Denis
Encyclopédie 161
Supplément au voyage de Bougainville 161
Dinawari, Abu Hanifah,
Kitab al-Nabat 38
Dionysius 26
Dioscorides Pedanius, *26*, *42*, 43, 44, 60, 64, 74, 76, 111, 115
De materia medica 23, 26, *27*, 33, 38, *40*, 61, 66
Mattioli's translation of 76
Dodoens, Rembert,
Cruydeboeck 163
Douin, Robert 224
Drake, Sarah Anne 196
Dransfield, John,
Genera Palmarum: The Evolution and Classification of Palms 236
Dridhabala 19
du Bois, Charles 126
Dürer, Albrecht 64
The Large Piece of Turf 67
Lily 66
Dutch East India Company (VOC) 113, 115, 116, *117*, 128, 134
Dutch West India Company (WIC) 115

E

East India Company 113, 126, 163, 170
Ebers Papyrus 14–16, *15*
Edwards, Sydenham *231*
Egyptians, ancient 14–16, 21
Ehret, Georg Dionysius 126, *136*
EIC (British East India Company) 250, 254
Elías, Baltazar 113
Ellis, Alice Blanche 218
Enlightenment 140
Erebus 161
Esau, Katherine,
Plant Anatomy 254–5
Euro+Med PlantBase 239
Europe 43–9, 128–30
Evans, Alice,
The Herefordshire Pomona 224
Evert, Ray F. 254–5
Ex herbis femininis 26, 33

F

Fauconberg, Lord 68
ferns 109, 198, *208*, 210, 212
Ferrari, Giovanni Battista,
De florum cultura 130
Fischer, Emil 186
Fitch, Walter Hood 198
Flora Antarctica 161
Fitter, R.S.R., *Collins Pocket Guide to Wild Flowers* 255
Fletcher, Henry 101
Flinders, Captain Matthew 160
Flora of Australia 239
Flora Europaea 239, *239*
Flora of Macaronesia 239
Flora of North America 239
Flora of the People's Republic of China 240–1
Flora Reipublicae Popularis Sinicae 239
Flora of Tropical East Africa 239
Flora of the USSR 239
Flora of West Tropical Africa 239
florilegium 88, 91
The Florilegium: The Royal Botanic Gardens Sydney: Celebrating 200 Years 250, *250*
A Florilegium: Sheffield's Hidden Garden 250
Florilegium amplissimum et selectissimum 101
Florilegium Society 250
Flowering Plants 233
Forster, George 141, *147*
Characteres generum plantarum 141
Leptosporum collinum 144
Forster, Johann 141, *147*
Characteres generum plantarum 141
France 151–4, 176, 224
Francis, George William,
An Analysis of the British Ferns and their Allies 212
Francis I 156
Franklin, Rosalind 230
Frontiers in Plant Science *249*, 250
Fuchs, Leonhart 66–73, *68*, 76, 154
Codex Fuchs 66, 68
De historia stirpium commentarii insignes 66, 68, *69–73*
Errata recentiorum medicorum 66
Neue Kreüterbuch 73
Füllmaurer, Heinrich 66, *68*
Furber, Robert,
Twelve Months of Flowers 101, *102–5*
Fust, Johann 61

G

Galen 43, 60
De simplicium medicamentorum temperamentis ac facultatibus 23
Simples 38
Gauci, Maxim 170

Gemmingen,
Prince-Bishop Johann Konrad 88
Genera Palmarum: The Evolution and Classification of Palms 236, *236*
Geological Survey 198
George, Alex 236
The genus Banksia L. f. (Proteaceae) 236
Gerard, John 77, 81
The Herball, or Generall Historie of Plantes 68, *79*, 81
Germany 154–6
Ghini, Luca 66, 74–5, *74*, 76
hortus siccus 75
'Global Assessment Report on Biodiversity and Ecosystem Services' 247
Gmelin, Johann 154
Flora Sibirica 154
Gorachand 170
Gosling, Raymond 230
Govaerts, Rafaël 245
Graff, Anton *147*
Greco-Roman culture 22–5
Greeks 20–2
Grew, Nehemiah 130, *132*
The Anatomy of Plants 130, *132–3*
Grimani Breviary 60, 63
Grindon, Leo,
Manchester Flora 201
Gronovius, Jan Frederik 134
Gutenberg, Johannes 52, 61

H
Harris, Stephen 77
Hartog, Jan 116
Hearst Papyrus 16
Hennig, Willi 225, *227*
Phylogenetic Systematics 234
Henry VIII 64
Henslow, John Stevens 196, 210
Hermann, Paul 116, *117*
Hernández, Francisco 52, *112*, 115
Rerum medicarum Novae Hispaniae thesaurus 113
Heuresis *26*
The Highgrove Florilegium 250
Hilder, Rowland and Edith,
Shell Guide to Flowers of the Countryside 262
Hill, Sir John,
The Vegetable System 151
Hill, Robin 186
Hill-McEntee, Susan 250, *252*
Hinduism, *Vedas* 16
Hippocrates 20, 60
Hippocratic Corpus 20, *20*, 21
Hist, Johann and Conrad 62
Historia naturalis 23
Hisui Sugiura,
Album of 100 Flowers 190–1
Hodgkin, Thomas 192
Hogg, Dr Robert
British Pomology 218
The Herefordshire Pomona 218, 224
Holbein, Hans 64
Hooke, Robert *131*
Micrographia 130, *131*
Hooker, Harriet 198
Hooker, Joseph
161, 198, *198*, 201, 210, 233
Flora Antarctica 161
Flora of India 198
Genera plantarum 210, *211*
Himalayan Journals 202
Illustrations of Himalayan Plants 198, *200*
Rhododendrons of Sikkim-Himalaya 198
Hooker, William Jackson
177–83, *178*, 198
A Century of Ferns 212
Curtis' Botanical Magazine 179
Exotic Flora 182–3
Flora Scotia 178
Muscologia Britannica 177, *179*
Plantarum 180–1
The Horae beatae Mariae ad usum Romanum 60
Horsefield, John 201
Humboldt, Alexander von
141, 162–3, *174*
Essai sur la géographie des plantes 162
Hunayn bin Ishaq Ibadi 38

I
Ibn al-Baytar *38*, 39, *39*, *42*, 43, 81
The Book of Medicinal and Nutritional Terms 41
Kitab al-Jami' li Mufradat al-Adwiyah w-al-Aghdhiyah 38, 43
Ibn Basilos, Staphen 38
Ibn Sina 38, *39*, *42*, 43
Icones Plantarum Malabaricum 122
Ieyasu, Tokugawa 127
Index Kewensis 245, *246*
Index Plantarum Japonicarum 226
India 16–19, 21, 37
Indus valley 14, 21
Intergovernmental Panel on Climate Change (IPCC) 243–4
IPCC Assessments 244
International Plant Names Index 245
International Union for the Conservation of Nature (IUCN) 247
Isingrin, Michael 66
Islamic plant science 37–43
Italian botanic gardens 74
Iwasaki Tsunemasa 186, 192
Honzō Zufu 186, 187, *187*, *188–9*, 192
Somoku sodate-gusa 187

J
Jacquin, Nikolaus von 156–7
Florae Austriacae 156
Hortus botanicus vindobonensis 156
Plantarum rariorum 156
Selectarum stirpium americanarum 156, *157*
Janssen, Hans 130
Janssen, Zacharias 130
Japan 126–30, 187–92, 225
Jardin des Plantes (Jardin du Roi), Paris 151–4, 170, 176
Jekyll, Gertrude 214, *214*
Wood and Garden 214
Jensen, Nicolas 54–5
Johann of Speyer *53*, 54–5
John Reeves Collection, Natural History Museum *237*
Josephine Bonaparte
170, *170*, 176
Jussieu, Antoine Laurent de
176–7, 178
Genera plantarum 176

K
Kaempfer, Engelbert:
Amoenitatum exoticarum 128, *128*
Flora Japonica 128, 130, 137
Kandel, David 76–7
Kangxi, Emperor 126
Keble Martin, W.,
The Concise British Flora in Colour 255
Keiga, Kawahara 187
Kesseler, Rob,
Seeds: Time Capsules of Life 254, *254*
Kimmerer, Robin Wall,
Braiding Sweetgrass: Indigenous Wisdom, Scientific Knowledge and the Teaching of Plants 262
Knopp, Johann Hermann,
Pomologia 169
Kolbert, Elizabeth,
The Sixth Extinction: An unnatural history 262
Kölreuter, Joseph 154
Krateuas 26
Kühne, Wilhelm 186
Kunth, Carl *174*

L
Laet, Johannes de 115
Lamarck, Jean-Baptiste
151, 154, 177
Flore Française 154
Lambert, Aylmer,
A Description of the genus Pinus 167, 168
Lankester, Phoebe 210
A Plain and Easy Account of the British Ferns 210
Lawrence, Ernest 224, *227*
Le Blon-Gauthier method 151
Le Moyne, Jacques 83
Grapes on the vine 85
map of Florida *84*
l'Écluse,
Charles de (Carolus Clusius) 77
Lee, James 140
An Introduction to Botany 140, *140*
Lee, James Jr 140
Leopold, Aldo 216, *217*
A Sand Country Almanac 217
L'Héritier de Brutelle,
Charles Louis 170
Li Shizhen,
Ben Cao Gang Mu 126–7
Liebig, Justus von, *Die organische*

Chemie in ihrer Anwendung auf Agricultur und Physiologie 186
Lightfoot, John, *The Flora Scotica, or a systematic arrangement, by the Linnean Method, of the native plants of Scotland and the Hebrides* 148–9
Lignamine, Johannes Philippus de 61
Lindley, John 157
Illustrations of Orchidaceous Plants 193
Ladies' Botany 201
Linnaean system 151, *151*, 157, 160, 176, 177, 178, 186, 192
Linnaeus, Carl 9, 10, 116, 134–7, *134*, 140, 141, *148–9*, 154
Arabidopsis thaliana 232
Bibliotheca botanica 136
Classes plantarum 136
Critica botanica 136, 137
Flora Lapponica 134
Fundamenta botanica 136, 137
Genera plantarum 136, *136*, 137
Hortus Cliffortianus 136
Oraeludia sponsaliorum plantarum 135
Philosophia botanica 136, 137
Species Plantarum 137
Systema naturae 134, *136*, 137
Linnean Society 10, 149, 151, 157, 177, 192, 198, 214
Lister, Joseph Jackson 192
l'Obel, Matthias (Lobelius) 81
A New Notebook of Plants 81
Loudon, Jane *204*
Botany for Ladies 201, *201*
Ladies Flower Garden Annuals 204
Ladies Flower Garden of Ornamental Perennials 205–7
Loudon, John *204*
Lovelock, James
A New Look at Life on Earth 262
Lutyens, Edwin *214*

M
McAlpine, Daniel 210
Botanical Atlas 210
McClintock, David,
Collins Pocket Guide to Wild Flowers 255
Macer Floridus,
De viribus herbarum carmen 60–1
Magazine of Botany 165
Magnol, Pierre, *Prodromus historiae generalis, in qua familiae per tabulas disponuntur* 111
Majerus, Marianne 255, *259*
Gardens of Corfu 255, *255*
Gardens of the Italian Lakes 255
The Great Gardens of London 255
Makino, Tomitaro *192*
Marcgrave, Georg 115
Historia naturalis Brasiliae 114, 115
Marrell, Jacob 122
Marshal, Alexander,
Mr Marshal's Flower Book 91, *94–9*
Masarjawayh 38
Masson, Francis 140
Matsumura, Jinzō 225
Nippon Shokubutsumeii 226
Mattioli, Pietro Andrea 66, 76
Commentarii in sex libros Pedacii Dioscoridis Anazarbei de Materia medica 76
Maurits, Count Johan 115
Maximilian II 88
Mayer, Albrecht 66
Med-Checklist 239
Medici, Cosimo de 74
Medieval period 26–34, 37
Mee, Margaret,
In Search of Flowers of the Amazon Rainforest 262, *262*
Megenberg, Konrad von,
Das Buch der Natur 55, *55*, *58*, 60
Mendel, Gregor 212–13, *213*
Mendoza, Francisco de 83
Merian, Maria Sibylla,
Metamorphosis Insectorum Surinamensium 122, *123*
Merrick, Emily *210*
Mesopotamia 14
Metrodorus 26
Mexico 81–3, *112*, 113
Miescher, Friedrich 230
Millennium Ecosystem Assessment 243
Miller, J F, African corn lily *143*
Miller, John,
An Illustration of the Sexual System of Linnaeus 148–9
Mirbel, Charles,
Traité d'anatomie et de physiologie végétale 192
Mohammad 37, 38
Mohl, Hugo von 186
Moldenhawer, Johann 192
Møller, Ian Max,
Plant Physiology and Development 255
Monson, Lady Anne 140, *141*
Morocco 81
Mortimer, Cromwell 126
Muir, John 218, *218*
My First Summer in the Sierra 218
Murphy, Angus,
Plant Physiology and Development 255
Museum of Natural History, Paris 176
Mutis, José Celestino 161–2, *161*, 163

N
Nagoya Protocol 244
Nalanda, Bihar 6
Napoleon Bonaparte 176
National History Museum, London 141
National Parks Act (1949) 216
National Trust 218
Nature 230
Nature Conservancy 216
Neleus 6
Newman, Edward,
A History of British Ferns 212
Nicander of Colophon 22
Alexipharmaka 22
Georgica 22
Nicolaus of Damascus,
De plantis 21, 38, 43
Niel, Cornelis van 186
Nodder, Frederick Polydore,
Hibiscus meraukensis 146
Noltie, Henry 254
Northern California Society of Botanical Artists 250

O
Observer, *British Grasses, Sedges and Rushes* 224
Odum, Eugene Pleasants
Ecology 255
Ecology and our Endangered Life Systems 255
Oldenland, Heinrich Bernhard 116
Olybrius, Flavius Anicius 26
Ono Ranzan 163, 192
Ben Cao Gang Mu 192
Honzō kōmoku keimō 163
Oppenheim Quran *39*
orchids 193–6, 197, *197*, 247
Orta, Garcia d', *Colóquios dos simples e drogas da Índia* 83–4

P
Padua, Italy 74
Pandit, Vinayak 117
Parasara, *Vrksāyurvēda* 37
Parkinson, John 91, *91*
Paradisi in sole Paradisus Terrestris 90, 91, *91–3*, 100
Theatrum botanicum 106
Parkinson, Sydney 141, *142*, *143*, 146
Metrosideros collina 145
Paulli, Simon,
Flora Danica 106, *107*
Paxton, Joseph,
Magazine of Botany 165
Petri, Johann, *Passau Herbarius* 62
Petrides, George A.,
A Field Guide to Trees and Shrubs 255, *261*, 262
Petru, Sally 250, *252*
Pfister, Albrecht 55
Philip II, King of Spain 113
Pisa 74
Piso, Willem 115
Plant Patent Act (1930) 213
Plant Physiology and Biochemistry 230
Plantin, Christopher 63, 81
Plants, People, Planet 230
Plants of the World Online 247
Platearius, Matthaeus,
Circa Instans 44
Plato 21
Pliny the Elder (Gaius Plinius Secundus) *25*, 26, 64, 111
Historia naturalis 23–5, *24*, 38, *53*, 54–5
Medicina Plinii 25
Pliny the Younger 23, 25, 54

The Pocket Guide to Wild Flowers *260*
Porro, Girolamo 74
Potter, Beatrix 9–10, 218, *219*
mycology *219, 220*
Prasad, Vishnu 170
Pseudo-Apuleius' *Herbarius* 25, 26, *32*, 61

Q
the Quran 37–8, *39*

R
Ramayana 18
al-Rashid, Harun 38
Ray, John 106, *106*, 109, 111, 137, 176
Catalogus plantarum Angliae *108*, 109, *109*
Catalogus plantarum circa Cantabrigiam 109
Historia plantarum *110*, 111
Methodus plantarum nova 111
al-Razi, Abu Bakr 38
Recchi, Antonio,
Nova Plantarum animalium et mineralium Mexicanorum historia *112*
Red List of Threatened Species 247
Redouté, Pierre-Joseph 170–6
Choix Des Plus Belles Fleurs *172, 174–5*
Les Liliacée 173
Les Roses 171, 176
Peintre des Fleurs 170
Reflora Virtual Herbarium *242*, 243
Regnault, Geneviève de Nangis,
La Botanique Mise à la Portée de Tout le Monde 151, *152–3*
Regnault, Nicolas Francois,
La Botanique Mise à la Portée de Tout le Monde 151, *152–3*
Renaissance 11, 54, 60, 63, 74, *79*
Reynolds, Sir Joshua *142*
Rheede, Hendrik van 116
Hortus Malabaricus 116–17, *118–21*, 122, 137
Rigveda 16–17, *17*
Rix, Martyn,
Indian Botanical Art: An Illustrated History 254
Robinson, William 214, *214*
Alpine Flower for Gardens: Rock, Wall, Marsh Plants, and Mountain Shrubs 214
The Garden 214
Hibiscus archeri 215
The Wild Garden 214
Rosser, Celia,
The Banksias 236, 239
Royal Botanic Gardens, Kew 141, 160, 163, *164*, 170, *178*, *180*, 198, 210, 224
Shirley Sherwood Gallery 250
'State of the World's Plants and Fungi' (SOTWPF) 244–5, *245*, 247
Victoria boliviana 248
Royal Botanic Garden Edinburgh 254
Royal Botanic Gardens, Sydney, Australia 250, *250*
Royal Botanical Expedition to New Granada 161–2
Royal Horticultural Society (RHS) 193, 198
Royal Society 130, 141
Royal Swedish Academy of Sciences 136
Rozzo, Stephanie 250, *253*
Rudbeck, Olaf 134, *137*
Rumphius, Georg Eberhard 137
Herbarium amboinense 115–16, *116*
Rumphius, Paul August 116
Rumphius, Susanna 116

S
Sahagún, Fray Bernardino de 83
Florentine Codex 82, 83, *83*
Salerno, Italy 43–4
Samaveda 16
Saussure, Nicolas de 186
Scheidel, Franz von 156
Schleiden, Matthias 192–3, *192*
Beiträge zur Phytogenesis 193
Schöffer, Peter 61–3
Herbarius latinus 61–2
Ortus Sanitatis *56–7, 59*, 62–3, *62*
Passau Herbal 62
Schott, Johann 64
Schwann, Theodor 192–3
Mikroskopie 193
Seeds: Time Capsules of Life 254, *254*
SEIC (Swedish East India Company) 136
Serapion the Younger,
Carrara Herbal 44, *45–9*
Sextius Niger 25
Shen Nong *34*
Shen Nong Ben Cao Jing 34–5
Sherard, William 126
Musaeum Zeylanicum 116
Sherwood, Shirley 250
The Shirley Sherwood Collection 250, *251*
Sibthorp, John 157
Flora Graeca 157, *158–9*
Siebold, Philipp von 187
Flora Japonica 187
Siegesbeck, Johann 134, 136
Sivry, Philippe de 88
Sixtus the IV, Pope 61
Sloane, Sir Hans 68, 126
Smith, James Edward 157
English Botany 149, 151
Smith, Lucy T. 248, 250
Smith, Matilda 198, *198*
Titan Arum 199
Smith, William 141
Sneath, Peter,
Numerical Taxonomy 233–4
Sokal, Robert,
Numerical Taxonomy R. 234
Solander, Daniel 137, 140, 141
South America 81–3, 161, 162
Sowerby, James 149
Speckle, Vitus Rudolph 66
Sprengel, Christian 154, 156
Das entdeckte Geheimnis der Naturi m Bau und in der Befruchtung der Blumen 156
'State of the World's Plants and Fungi' (SOTWPF) 244–5, *245*, 247
Strabo 6
Stuppy, Wolfgang,
Seeds: Time Capsules of Life 254, *254*
Surupala, Vrksāyurvēda 37
Sushruta Samhita 19, *19*
Swan, J. *182–3*
Swedish East India Company (SEIC) 136
Sweert, Emmanuel 101

T
Tahiti 161
Taiz, Lincoln,
Plant Physiology 255
Takhtajan, Armen 233, *234*
Diversity and Classification of Flowering Plants 233, *234*
Systema Magnoliophytorum 233
Tang Shenwei,
Zheng Lei Ben Cao 37
Tansley, Arthur 214, 216, *216*
Tao Hongjing,
Ben Cao Jing Ji Zhu 34, 35–6
Tärnström, Christopher 136
Taylor, Thomas,
Muscologia Britannica 177, *179*
Taylor, William 198
Tenneson, Joyce 255
Flower Portraits: The Life Cycle of Beauty 255, *255*, *256–7*
Thabit bin Qurrah 38
Thal, Johannes 232
Sylva Hercynia 106, *232*
Theophrastus *21*, 25, 75, 111, 115
and Alexander the Great *22*
De causis plantarum 6, 21, 38
Historia Plantarum 6, *20*, 21–2, 23, 38
Thesaurus Linguae Japonicae 128
Thory, Claude-Antoine,
Peintre des Fleurs 170
Toledo, Spain 44, *44*
Tournefort, Joseph Pitton de 134, 176
Élémens de Botanique ou Méthode pour Connoître les Plantes *110*, 111
Tradescant, John the Elder 91, 101
Tradescant, John the Younger 91, *94*
Tradescant's Orchard 101, *101*
Traditional Chinese Medicine (TCM) 35
the Transylvania Florilegium 250, *251*
Trota of Salerno,
Practica secundum Trotam 44

tulips
76, 88, *88*, *92*, 100–1
Tulley, William 192
Turner, William 68
Tyler, John 170

U
University of Bologna 77, *77*
US National Parks 218, *218*

V
Varro, Marcus Terentius 25
Vaucher, Jean Pierre 177
Vázquez, Pedro 113
Vedas 16–17, 19
Veitch, James 214
Victoria boliviana
248, *248*, *249*
Virgil (Publius Vergilius Maro)
22, 25
Visnupersaud 254
VOC (Dutch East India
Company)
113, 115, 116, *117*, 128, 134
Vries, Hugo de *Die*
Mutationstheorie 213
Vrksāyurvēda 37

W
Wales, HRH The Prince of
250
Wallace, Alfred Russel
196, 214
Wallich, Nathaniel
163, *164*, *165*
Plantae Asiaticae Rariores
163, *164*, 170
Wang Haogu,
Tang ye Ben Cao
36, 37
Warming, Eugen 214
Oecology of Plants 216
Plantesamfund 214
Warne, Frederick,
British Wild Flowers 224
Watson, James 230
Weiditz, Hans 64, *65*
Wendelin of Speyer 54–5
WIC (Dutch West India
Company) 115
Wiley *231*
Willstätter, Richard 224, *227*
Wilson, E. H., 'Chinese'
The Cherries of Japan
225, *226*
Wilson, E. O.,
The Diversity of Life 255, 262
Withers, Augusta 196
'World Checklist of Vascular
Plants' 245, *246*, 247, *247*

X
Xin Xiu Ben Cao 36

Y
Yajurveda 16
Yonge, Charlotte Mary,
The Instructive Picturebook,
or Lessons from the Vegetable
World 209
Yosemite National Park 218

Z
Zainer, Günther 55
Zeiger, Eduardo,
Plant Physiology 255
Zeiss, Carl 193
Zimmermann, Walter 225
Zuccarini, Joseph Gerhard 187

FURTHER READING

Anwer, Dr.R., *Arab contribution to Botany and Agriculture.* New Delhi: Adam Publishers, India, 2016.

Attenborough, David, *Amazing Rare Things, the Art of Natural History in the Age of Discovery,* Yale University Press, Connecticut, 2007.

Bynum, Helen and William, *Botanical Sketchbooks,* Thames & Hudson, London, 2017.

Beatty, L., *Looking for Theophrastus: Travels in Search of a Lost Philosopher,* Atlantic Books, London, 2022.

Beil, Karen Magnuson, *What Linnaeus Saw: A Scientist's Quest to Name Every Living Thing,* W. W. Norton & Company, New York, 2019.

Bensky, Dan, *Chinese Herbal Medicine: Materia Medica,* Eastland Press Inc, Seattle, 1993.

Blunt, Wilfrid, *The Art of Botanical Illustration,* Collins, London, 1950.

Blunt, Wilfrid and Raphael, Sandra, *The Illustrated Herbal,* Frances Lincoln, London, 1979.

Bowler, Peter, *The Fontana History of the Environmental Sciences,* Fontana, London, 1992.

Coleman, William, *Biology in the Nineteenth Century,* Cambridge University Press, Cambridge, 1977.

Cuvier, Georges and Pietsch, Theodore Wells, *Cuvier's History of the Natural Sciences – Twenty-four lessons from Antiquity to the Renaissance,* Scientific Publications of the Muséum national d'Histoire naturelle, Paris, 2012.

Damodaran, V., Winterbottom, A. and Lester, A. (eds), *The East India Company and the Natural World,* Palgrave Macmillan UK, London, 2015.

Debus, Allen G, *Man and Nature in the Renaissance,* Cambridge University Press, Cambridge, 1978.

Drake, Ellen Tan, *Restless Genius: Robert Hooke and His Earthly Thoughts,* OUP USA, 1996.

Endersby, Jim, *Imperial Nature, Joseph Hooker and the Practices of Victorian Science,* University of Chicago Press, Chicago, 2008.

Endersby, Jim, *Orchid, A Cultural History,* University of Chicago Press, Chicago, 2016.

Fara, Patricia, *Sex, Botany and Empire: The Story of Carl Linnaeus and Joseph Banks.*: Icon Books, Cambridge, 2003.

Grove, Richard H, *Green Imperialism: Colonial Expansion, Tropical Island Edens and the Origins of Environmentalism, 1600–1860 (Studies in Environment and History),* Cambridge University Press, Cambridge, 2010.

Hardy, G. and Totelin, L. *Ancient Botany,* Taylor & Francis, London, 2015.

Harris, Stephen A., *The Beauty of the Flower, the Art and Science of Botanical Illustration,* Reaktion Books, Chicago, 2023

Heilmeyer, M., *Ancient Herbs,* Getty Publications, Los Angeles, 2007.

Henderson, L.D.J'Greek Wild Flowers and Plant Lore in Ancient Greece'. H. Baumann, translated and revised by W. T. & E.R. Steam. Herbert Press, London, 1993, *Edinburgh Journal of Botany,* 50(2), pp. 241–242.

Jardine, N., and Spary, E J. (eds), *Cultures of Natural History,* Cambridge University Press, Cambridge, 1996

Keyser, Paul T. and Irby-Massie, Georgia L., *Encyclopedia of Ancient Natural Scientists: The Greek Tradition and its Many Heirs,* Routledge & CRC Press, Oxford, 2008.

Lack, H. Walter, *A Garden Eden, Masterpieces of Botanical Illustration,* Taschen, London, 2016.

Leith-Ross, P., *John Tradescants, The: Gardeners to the Rose and Lily Queen,* Peter Owen, London, 2006.

Magee, Judith, *Rare Treasures from the Library of the Natural History Museum,* The Natural History Museum, London, 2017.

Manilal, K.S., *Hortus Malabaricus and the socio-cultural heritage of India,* Wellcome Collection, London, 2012.

Muir, John, *Nature Writings,* Library of Congress, Washington D.C., 1997.

Nagar, S.L. B*otanical and Medicinal Plants as Depicted in Ancient Texts Art.* New Delhi: B.R. Publishing Corporation, 2000.

Nair, S.R. and G.S.U., *Vrikshayurveda – Ancient Science of Plant Life and Plant Care,* Thiruvananthapuram: Kerala State Biodiversity Board, Kerala, 2017.

Nutton, Vivian, *Ancient Medicine,* Routledge & CRC Press, Oxford, 2023.

Parkinson, Anna, *Nature's Alchemist: John Parkinson – Herbalist to Charles I,* Frances Lincoln, London, 2007.

Pavord, Anna, *The Naming of Names,* Bloomsbury, London, 2007

Pickstone, John V, *Ways of Knowing, A New History of Science Technology and Medicine,* Manchester University Press, Manchester, 2000.

Singh, A. *Plants in Ancient Indian Civilization,* Agam Kala Prakashan, India, 2008.

Turner, G.L., *Essays on the History of the Microscope,* Senecio Pub. Co., Oxford, 1980.

PICTURE CREDITS

The publishers thank the following for permission to reproduce the illustrations in this book. Every effort has been made to provide correct attributions. Any inadvertent errors or omissions will be corrected in subsequent editions.

Cover: GRANGER - Historical Picture Archive / Alamy Stock Photo
Back cover: Biodiversity Heritage Library

Alamy Stock Photo: 8, 20 right + 120–21 (Volgi archive), 15, 77 left, 106 bottom left, 133 + 218 (Science History Images), 39 left (Peter Horree), 39 bottom right (AF Fotografie), 45 (PBL Collection), 46–7, 48–9, 81, 95, 96, 116, 160, 177 bottom, 215, 216 top left + 224 (The Picture Art Collection), 62 left (Charles Walker Collection), 84–5 + 180– 81 (GRANGER - Historical Picture Archive), 40, 90, 173, 179 left, 203 bottom right + 207 (Penta Springs Limited), 91 bottom (World History Archive), 94 top left (Heritage Image Partnership Ltd), 102 (imageBROKER.com GmbH & Co. KG), 104–5, 171, 182, 183, 189, 206 + 223 (Album), 108 (Historic Images), 118–9 (Chronicle), 128 (Zoom Historical), 129 + 225 bottom (Signal Photos), 142 right + 217 top left (IanDagnall Computing), 144 + 225 top (Yogi Black), 147 right + 192 bottom (Sueddeutsche Zeitung Photo), 163 bottom (PRISMA ARCHIVO), 136 left, 164 top left, 170, 178, 198 top + 204 right (The Natural History Museum), 164 midde left, 193 + 210 (The History Collection), 164 right + 174 (Historic Illustrations), 165, 188 + 205 (Florilegius), 166 left (BTEU/ AUSMUM), 166 right (INTERFOTO), 172 (The Nature Notes), 175 (Artepics), 177 top + 227 left (Pictorial Press Ltd), 194 + 195 (Tibbut Archive), 199 (Alto Vintage Images), 202 (History and Art Collection), 203 top left (MET/ BOT), 203 top right (Natural Visions), 203 bottom left (Axis Images), 204 top left (Zuri Swimmer), 208 (19th era), 209 (Marcus Harrison – botanicals), 214 bottom (De Luan), 217 top right (sjbooks), 219 top right (WorldPhotos), 226 top (Eraza Collection), 248 (Guy Bell), 251 top (Monica Wells), 257 top (Robert Landau), 261 top left (NiaBell)

Biodiversity Heritage Library: 2, 106 top left + middle, 109 left + right, 117 left, 127 right, 158 right, 159, 162, 167, 168, 169, 200, 211, 213 right, 216 bottom left + bottom right, 226 bottom left, 231, 243

Bodleian Library, Oxford: 101

Brazil Reflora: 242

Bridgeman: 7 (British Library archive), 9 + 82 (© NPL - DeA Picture Library), 24 (© Musée Condé, Chantilly), 27, 145, 146 left + 236 (© Natural History Museum, London), 66, 83 top + 146–47 (Bridgeman images), 83 bottom right (© Raffaello Bencini), 91 top (Universal History Archive/UIG), 97 (Royal Collection Trust / © His Majesty King Charles III, 2024), 152 (© Archives Charmet), 153 (Photo © CCI), 155 (© Florilegius), 197 (Art Collection 3), 217 bottom (LAYNE KENNEDY / Contributor), 256 (Saint Louis Art Museum / Friends Fund)

Bulfinch Press: 253, 255 top

Cambridge University Press: 244 left

Columbia University Press: 234 right, 235 left, 239

Connoisseur: 240–41

Los Angeles County Museum of Art: 19

Digital Public Library of America: 226 bottom middle

Frontiers: 249 top

Impress: 255 bottom

Getty Images: 28–9 + 30–1 (Science & Society Picture Library / Contributor), 43 (Art Images), 103 (Historical Picture Archive), 214 top (English Heritage/Heritage Image), 221 + 22 (Florilegius/Universal Images Group), 238 (Fairfax Media/ The AGE / Contributor, 257 bottom (Anthony Barboza / Contributor)

HathiTrust: 10, 148-49, 187 left + right, 226 bottom right

Houghton Mifflin: 261 bottom left + bottom right

Library of Congress: 36, 55 left + right, 157

Lucy Smith: 249

Marianne Majerus Garden Images: 258 (© MMGI / Marianne Majerus / Sussex Prairies) + 259 (© MMGI / Marianne Majerus / Sleightholmedale Lodge, Yorkshire) + 260

The Mertz Library of the New York Botanical Garden: 235 right

Papadakis Publishers: 254

Phoenix House Ltd: 261 top right

Project Gutenberg: 201

Public Domain: 20 left, 21 bottom, 22, 26, 32–3, 35, 41, 42, 53, 56–7, 58–9, 60, 64, 65 top + bottom, 67, 68–9 , 70–1, 72–3, 74, 89, 106 r, 112, 114, 117 right, 124, 125, 126, 127 left, 131, 134, 135, 136 right, 143 left, 150, 151, 158 left, 192 top, 196, 198 bottom

Raby Estates: 141 top

Royal Botanic Gardens, Kew: 237, 245, 246, 247, 251 bottom

Royal Collection Trust: 94 right, 98, 99, 123, 140

Science and Society Picture Library: 143 right (© Science Museum Library / Science & Society Picture Library – All rights reserved)

Sistema Museale di Ateneo, Università degli Studi di Firenze UNIF: 77 right, 78

Smithsonian Libraries and Archives: 141 bottom

Southern Illinois University: 244

The Armitt Trust: 219 top left + bottom, 220 top + bottom

The Lavenberg Collection of Japanese Prints: 190 + 191

University of Chicago Press: 250

University of Oxford: 232 left + right

Wellcome Collection: 18, 25, 34, 44, 79 left + right, 80, 88 top + bottom, 161, 213 left + middle

Wikimedia commons:16, 17, 21 top, 38, 62 right, 107, 110 left, 136 middle, 137, 142 left, 163 top, 179 right, 227 right

ACKNOWLEDGEMENTS

Carolyn Fry and Emma Wayland would like to thank: Anne Marshall and other library staff at the Royal Botanic Gardens, Kew, UK, for their help with locating research material; and Richard Green, Jennifer Barr, Michael Brunstrom and all at Quarto Publishing for their hard work to bring this handsome book to fruition.

Carolyn Fry would also like to thank Alex Benwell for patiently listening to and commenting on early drafts of the chapter texts.

Emma Wayland would also like to thank Jon Howes, Kester Hall, Jenn Ashworth and Tristan Burke.